EARTH SCIENCES IN THE 21ST CENTURY

LIGHTNING: PROPERTIES, FORMATION AND TYPES

EARTH SCIENCES IN THE 21ST CENTURY

Additional books in this series can be found on Nova's website under the Series tab.

Additional E-books in this series can be found on Nova's website under the E-books tab.

SPACE SCIENCE, EXPLORATION AND POLICIES

Additional books in this series can be found on Nova's website under the Series tab.

Additional E-books in this series can be found on Nova's website under the E-books tab.

EARTH SCIENCES IN THE 21ST CENTURY

LIGHTNING: PROPERTIES, FORMATION AND TYPES

MATTHEW D. WOOD
EDITOR

Nova Science Publishers, Inc.
New York

LIBRARY OF CONGRESS CATALOGING-IN-PUBLICATION DATA

Lightning : properties, formation and types / editor, Matthew D. Wood.
p. cm.
Includes index.
ISBN 978-1-61122-198-5 (hardcover)
1. Lightning. I. Wood, Matthew D.
QC966.L534 2010
551.56'32--dc22
2010041313

Published by Nova Science Publishers, Inc. † New York

CONTENTS

PREFACE

Lightning balls are very rare natural phenomena that appear as free floating glowing globs, usually during electric activity in the atmosphere of the Earth. Each year, hundreds of eyewitness accounts of such strange events are reported to local journals, police and even military authorities. This book also discusses, extraterrestrial lightning detected so far on the planets Jupiter, Saturn, Uranus, Neptune, and Venus, which are attributed to intracloud lightning. An overview of the results of Schumann resonance and ELF studies obtained during the last decade are explored and educational computer simulations of lightning and associated thunders is reviewed.

Chapter 1 - Lightning balls are very rare natural phenomena that appear as free floating glowing globs, usually during electric activity in the atmosphere of the Earth. Each year, hundreds of eyewitness accounts of such strange events are reported to local journals, police and even military authorities.

Chapter 2 - All extraterrestrial lightning detected so far on the planets Jupiter, Saturn, Uranus, Neptune, and potentially at Venus is attributed to intracloud lightning. On Mars other types of discharges might exist within dust storms. In this commentary the authors will focus on the historical development of planetary lightning investigations at Jupiter, Saturn, and Venus. Special attention will be devoted to the controversies around the existence of lightning on Venus. Thereby the authors will highlight the importance of combined optical and radio observations for the epistemological degree of confidence in lightning detection from space as well as for its detailed investigation. Future spacecraft missions to various planets should take this into account. The main future extraterrestrial lightning investigations will be outlined. The authors also plead for radio observations with a time resolution of the order of microseconds and for more modelling and laboratory work in this field.

Chapter 3 – The authors overview the results of Schumann resonance and ELF studies obtained during the last decade. The following items are addressed: ionospheric non-uniformities related to seismic activity, formal proofs of the resonance line splitting under the influence of geomagnetic field, impacts of solar proton events and extra-solar gamma-ray bursts on resonance parameters, ELF intensity and global land temperature, Schumann resonance on other planets, data from a new observatory at Spitzbergen, sprites and ELF transients, biological effects of Schemann resonance, solutions of the inverse problem, the 'terminator effect' and its controversial explanations, the accuracy of locating the Q-bursts sources, a comparison of waveforms of Q-bursts (experiment versus theory), parameters of thunderstorm activity which follow from the optical transient detector data.

Chapter 4 - Lightning and thunder have been observed since a long time in the history of man. Speculations on the origin and cause of thunder date to the ancient Greece, with the opinion of Aristotle, relating them to sudden ejections of vapours from clouds. Lucretius in his "*De rerum natura*" subsequently attributed the thunder to the collisions between clouds. During the next centuries, the explanations concerning the cause of thunder made no reference to the existence of lightning in the sky. At the beginning of 1700, W. Wall and I. Newton commented that the origin of lightning could be similar to that of crackling of sparks produced by rubbed amber rods. Benjamin Franklin, in 1749, was probably the first one to suggest the electrical nature of thunderstorms, and to carry out experimental investigations aiming at demonstrating that thunderclouds were charged, and that a large potential difference existed between clouds and the Earth.

Chapter 5 - Why are some astrophysicists concerned with lightning? Because the current understanding of the planet formation theory has several problems, such as inconsistency, and missing pieces of the puzzle. By the way, although the effect of electric charge of the dust aggregates has been strangely neglected by most astrophysicists until recently, it has actually been found to be a crucial factor. If the authors consider the global motion of such charged dust aggregates, the generated electrostatic field is sometimes large enough to cause lightning.

To share their current understandings and problems with you, the authors first introduce you to the theory of planet formation, which takes place in protoplanetary disks.

Chapter 6 - *Ball Lightning: An Unsolved Problem in Atmospheric Physics* is the title of the last monograph published in 1999. The authors would like to refine this assertion and declare that Ball Lightning (BL) is a shame of physics today. At present more than 2000 papers and reports have been published, more than 200 various BL theories linking BL to a variety of physical, chemical and nuclear processes have been proposed but none of these theories seems to have gained a general recognition because they fail to explain all observed characteristics of the phenomenon. A regular appearance of new theories confirms this conclusion, and new theories are no exception. Systematization and classification of theories have been carried out. However neither of them even approximately can explain enigmatic and intriguing BL behaviour, which to a certain degree reminds us of the behaviour of some highly organized matter. Physicists cannot imagine the object, governed by conventional laws of physics, properties of which coincide with BL properties. Explanation of Ball Lightning physics is deadlocked.

Chapter 7 - Brazil is the largest country in the tropical region and it is believed to have the largest cloud-to-ground (CG) lightning activity in the world, estimated from thunderstorm days and satellite data in about 50 millions CG lightning flashes every year. The first systematic observations of CG lightning in Brazil were done in the decade of 1960 through the record of the number of thunderstorm days at different sites. Later on observations of CG flashes were done by a network of flash counters, an instrumented tower and a clod-to-ground lightning location network, most of them in the Southeast region of the country. During the last two decades the research has increased considerably, expanding to other regions in the country and including the technique of high-speed cameras. The lightning location network has also expanded considerably from four sensors in 1988 to 47 sensors at present time, becoming the largest network in the tropics and the third largest national network in the world. In this overview, the main results obtained by these techniques are described and compared with observations in other locations in the world.

They include approximately 100 direct current observations, 500 optical observations by high-speed cameras and more than 20 million electromagnetic field observations by lightning location technique. As a whole, they can be considered as the largest existing database on lightning in the tropical region.

In: Lightning: Properties, Formation and Types
Editor: Matthew D. Wood
ISBN: 978-1-61122-198-5

Chapter 1

DYNAMICAL SELF-ORGANIZATION ACTING AS PHYSICAL BASIS OF THE BALL LIGHTNING PHENOMENON

E. Lozneanu and M. Sanduloviciu
Faculty of Physics, University AI. I. Cuza, Isai, Romania

INTRODUCTION

Lightning balls are very rare natural phenomena that appear as free floating glowing globs, usually during electric activity in the atmosphere of the Earth. Each year, hundreds of eyewitness accounts of such strange events are reported to local journals, police and even military authorities.

The presence of lightning balls was well documented more than a hundred years ago in prestigious periodicals, as for example, *Nature*. Because of their random appearance, a direct experimental investigation was not possible up yet. Consequently, information concerning their luminous aspects and behaviour are mainly based on eyewitnesses. For example, in the United States, lightning balls have been observed by about 5% of the adult population at some times in their lives [1]. Among the prominent individuals who have observed lightning balls are *Niels Bohr* and *Victor Weisskopf.*

The literature of this phenomenon considered hitherto as an enigma of the Science comprises books entirely dedicated to this subject [2-4], review papers focussed on this topic [5-8], as well data banks. Periodically, new aspects concerning the ball lightning appearance and behaviour are discussed at international symposiums. So, in this year the ball lightning phenomenon will be analysed at the international symposium in Saint Petersburg, Russia. Appearing as glowing gaseous body, various plasma theories have been published but none of them account for the emergence and proprieties completely satisfactory. All of them run into problems of their stability and relative long lifetime.

Taking into account the very great number of papers describing the appearance and observational characteristics of the ball lightning, two essential questions prevail. First there is the question of what form of energy could be stored in well-localized relatively small volume of air (of the order of 10^{-3} to 10^5J/cm^3) and second, what mechanism explains its occasionally observed long life (up to 100s). There are two essential distinct opinions. One of them starts form the hypothesis that the generation and lifetime of the ball lightning can be explained by considering the nature of the material located in it during its generation by the linear lightning. So, different reactions ranging from chemical to nuclear ones initiated by the energy injected by the linear lightning were considered to be at the origin of the ball lightning. The other opinion relates the appearance and lifetime of the ball lightning to a local concentration of high-frequency electric field energy. Such phenomena are possible after interference of electromagnetic waves generated by linear lightning or other electrical phenomena occurred during thunderstorm in the atmosphere of the Earth. This opinion is argued by measurements showing that lightning flashes generate damped electromagnetic waves whose duration surpasses the time span of their occurrence.

Starting from the first hypothesis, *Kadomtsev* [9] considered the ball lightning as a gas accumulator, charged by ordinary lightning, where the accumulated chemical energy is released in the form of an electric discharge. *Turner* [7] considered the ball lightning to be a thermo-chemical heat pump powered by the electric field of a thunderstorm. A theory of ball lightning based on the electric field associated with current filaments originating from a point where linear lightning strikes the surface of the Earth was proposed by *Lowke* [10]. A new kind of power source of the ball lightning, considered by *Hubler* [11] to the most appropriate to elucidate the ball lightning phenomenon, was proposed by *Abrahamson* and *Dinniss* [12]. They consider at the origin of the ball lightning chemical energy stored in pure silicon, which is unstable to oxidation at high temperature.

Starting from the second hypothesis, firstly proposed by *Kapitsa* [13], *Jennison* [14] suggested that the ball lightning consists of electric field energy trapped in an air-evacuated cavity of approximately spherical shape with an ionized sheet separating it from the atmosphere. *Zheng* [15] considered ball lightning to be an enclosed plasma shell generated by short radio waves associated with ordinary lightning. In the same context, microwave experiments performed by *Othsuki and Oforuton* [16] proved the possibility for producing fireballs in air by using microwaves. *Yasui* [17] proposed a model related to this experiment. Another model proposed by *Handel and Leitner* [18] explains ball lightning as a plasma caviton feed by a large atmospheric maser. The caviton is considered to be a non-linear quasi-stationary electric field localized in a plasma configuration in oscillations. *Ranada and Trueba* [19] suggested that ball lightning could be formed from knots of the electric and magnetic components of an electromagnetic field. For explaining the occasionally reported very high energy density delivered by some ball lightning during their disintegration *Altschuler, House and Hilder* [20] suggest to consider the presence of nuclear reactions. More recently *P. A. Torchigin and A. V. Torchigin* [21] show that ball lightning can be a pure optical phenomenon where only intense light and compressed air interact.

The great interest concerning the possible implication of the phenomena at the origin of the ball lightning in fusion research, industrial plasma engineering, and military applications were evidenced in the comprehensive paper published by *Roth* [1].

As known, luminous gaseous flaming globes were observed long time ago in laboratories where the transport of current in rare gases was experimentally investigated. Thus, as reported by Wayne Strattman [22], Irving Langmuir, a prominent American scientist, and his collaborators have published in *Scientific American,* beginning with the second decade of the 19th century, photographs revealing that free floating flaming blobs with sizes in the order of centimeters emerge spontaneously in plasma devices. Briefly speaking, their device consisted of a glass tube, filled with Argon at very low pressures. Two heated, i.e., electron emitting tungsten filaments were placed in opposite sides in the tube. Positively biasing one of the filaments with respect to the other one (250 Volt), a gaseous conductor known as physical plasma was produced. Known as the fourth state of the matter, the plasma located in the glass tube contains besides neutral and excited atoms, free positive ions and electrons. Bringing the terminal of a high frequency coil near the tube, a free-floating flaming globe spontaneously emerges in the glass tube. The globe consists, as authors reported, of a central red core, with a diameter about one centimeter, which bears positive charge. A thin dark space surrounds this region and, beyond this, a bright yellow negative sheath of glowing gas that gradually disappears. Instead of this negative sheath an electric double layer that borders the flaming globe spontaneously appears. The authors explained the appearance and the visual aspect of the flaming globe observed in the glass tube considering ionization and excitations of tungsten atoms.

In spite of the fact that the presence of beautifully colored flaming globs bearing strong resemblances to lightning balls was reported in a great numbers of papers published up yet, these experimental results were considered up yet not able to offer information to explain the actual origin of the ball lightning. This is due to its appearance in very rare gases and the relatively small energy involved in its emergence. This state of the art was changed in the last time by new laboratory experiments showing that by very quick local injection of energy it is possible to produce in air, under controllable laboratory conditions, free floating flaming globs lifetime of which substantially surpasses the time span in which they emerged. Today the flaming globes that appear in plasmas that transport direct current are dubbed *fireballs* [23,24] whereas, those which appear in plasma created by high frequency electric field *plasmoids* [25,26].

In the following we will describe some of these experimental results but also other ones obtained in different devices were plasma is present.

EXPERIMENTS WITH RELEVANCE FOR THE BALL LIGHTNING

In a recently published paper, *V. Dikhtyar and E. Jerby* [27] shown that fireballs can be ejected from a molten hot spot in air by localized microwaves. The evolution from the molten hot spot created in this way into a microwave cavity comprises three phases emphasized in Fig 1 [28]. Thus, briefly speaking: (i) A well-located hot plasma is created by microwave drill by vaporization of the substrate material; (ii) A fire column is ejected from the well located hot plasma; (iii) The fireball detaches from the hot spot and floats near the cavity border. This experiment proves, in our opinion, the presence of a succession of events by which matter transits from disordered state (the hot plasma in non-equilibrium) into an ordered state (the fireball), in absence of an external driving force, i.e., by self-organization. As all complexities emerged by self-organization [29, 30], the fireball emerged in this way reveals a certain robustness continuing to glow for another 30-40 ms after the microwave source is turned off. Another important experimental result presented in the same paper is that the ignition of the fireball is accompanied by strong absorption of microwave power. These experimental results were explained by the authors considering at the origin of the fireball a slowing "burning" process of solid material particles as silicon, germanium, glass, ceramics, basalt etc, initiated by very quick injection of microwave energy. The authors consider that their experimental results are strong arguments in favor of the model of the ball lightning proposed by *Abrahamson and Deniss* [12].

Another experimental results, also frequently commented in the last time in the scientific literature in connection with the ball lightning phenomenon, were published by researches at the Humbold University, University Leipzig, Germany and Eindhoven University, The Netherlands [31]. In that experiment, fireballs of diameters between 10-20 centimetres (Fig 2) were produced by very short high voltage discharge (up to 60 amperes) in water. The fireballs continued to be visible about 300 millisecond after the current has decayed and the energy input was thus out off. The fireball glows very brightly although it appears to be rather cold. Thus, sheet of paper placed above them dues get lifted but it not catch fire.

The above briefly described experimental results obtained in air at normal pressures were related to the ball lightning phenomenon because the fireballs appear in an environment similar to that in which lightning balls usually appear. However, phenomena as, for example, the motion of the fireball against the wind and its penetration through windows without damage them, many times reported by the observers, are events not explainable by the above described experiments. In this context there are reports that lightning balls are able to trail beside aircrafts. Occasionally, these penetrate the windows of aircraft moving inside them before it disintegrates in absence of producing important physical injuries.

The experimental results are important in connection with the ball lightning phenomenon because they proved that, under controllable laboratory conditions, it is possible to create in air free floating fireballs. Their lifetime substantially surpasses the time span in which they appeared after very quick injection of an amount of energy that must surpass a critical value. However, these experiments are not able to offer an insight concerning the actual physical processes and the mechanism involved in their origin. This lack in understanding the fireball appearance can be completed, as we will show in the following, by investigating the circumstances under which coherent, relative stable and luminous gaseous bodies emerge in the plasma diodes [32,33]. The scheme of a plasma diode as that used in our laboratories is presented in Fig 3. It consists of a glass tube filled with argon at low pressure (about some mTorr) and a set of electrodes marked by major letters. The plasma was produced by a direct current gas discharge ignited between a heated tungsten filament marked **F** and a metal cylinder **M** bordered at its open side by the grid **G.** The dc power source $\mathbf{U_1}$ connected through the resistor $\mathbf{R_1}$ at **F** produces the direct current for heating the filament. Using the dc power source $\mathbf{U_2}$, whose negative pole was connected through the resistor $\mathbf{R_2}$ at **F** and its positive pole at **M**, it was created an electric field in which the electrons emitted by **F** are accelerated at energies sufficient to ionize and excite the argon atoms. The plasma created in this way penetrates the grid **G** reaching by diffusion the surface of the anode **A**. The anode was a tungsten wire 2mm in diameter laterally isolated by glass. For working as a diode, a dc power source $\mathbf{U_3}$ was connected through the resistor $\mathbf{R_3}$ with its positive pole at **A** and with its negative pole at **M** and implicitly **G.** A movable electron emitting probe (emissive probe) that penetrates **A** in axial direction was used to measure the potential in front of the anode. Another very fine filament electrode **E** positively biased by $\mathbf{U_4}$ was used to produce, when certain investigations was performed, an electric spark at the surface of **A**. An auxiliary device consisting of an oscillator that generates a high frequency electric field (about 100

MHZ) between the plate electrodes $\mathbf{E_1}$ and $\mathbf{E_2}$ was used to investigate the phenomena that appear in the diode when the plasma is subjected to the influence of this electric field.

Using the above briefly described experimental device it was possible to identify step by step the succession of phenomena that are the premises for initiating the spontaneous emergence of a coherent, stable and luminous gaseous body that maintains its electrical contact with the anode through a channel (Fig 4). Dubbed in the following as nearly spherical fireball, the phenomena at its origin can be identified when the energy injected in the plasma diode is gradually increased and decreased. Thus, maintaining constant the temperature of **F** and the voltage of $\mathbf{U_2}$, the typically static current (I) versus voltage (V) characteristic, shown in Fig 5, was obtained by gradually increasing and decreasing the voltage of $\mathbf{U_3}$. This characteristic reveals abrupt variation of the resistance of the gaseous conductor (the plasma) located between **G** and **A**, i.e., of the internal resistance of the plasma diode. So, in a first relative small voltage interval of the voltage of $\mathbf{U_3}$, the current I proportionally increases with the potential V of the anode **A** (branch **a-b**), i.e., Ohm's law is valid. Note that owing to the presence of $\mathbf{R_3}$ the potential of the anode V differs from the voltage of $\mathbf{U_3}$. Continuing to increase gradually the voltage of $\mathbf{U_3}$, the slope of the static $I(V)$-characteristic changes (branch **b-b'**). In this voltage interval, a luminous sheath appears attached to the anode **A**. Its appearance proves that, after acceleration in the electric field sustained by $\mathbf{U_3}$, the electrons obtain kinetic energies sufficient to produce atom excitations. Losing their kinetic energy, these electrons are thermalized so that we identified a net negative space charge by electric probe measurement at that distance from **A** where the excitation cross-section function suddenly increases. Since the exited atoms quickly ($\approx 10^{-8}$s) return to the ground state, this region emits light. Acting as a barrier for I, this net negative space charge evolves concomitant with the gradual increase of the voltage of $\mathbf{U_3}$. Its barrier effect grows concomitant with the increase of the anode potential V, fact revealed by the appearance of the branch **b-b'** in the $I(V)$-characteristic shown in Fig. 5. Further increasing the voltage of $\mathbf{U_3}$, the potential V reaches the value marked **b'** for which I begins to increase slowly because the first ionization processes appear in front of the anode. A critical value of the voltage of $\mathbf{U_3}$ is reached when the potential V of the anode reaches the value marked **c**. In this moment I abruptly increases (branch **c-d**) under constant value of the voltage of $\mathbf{U_3}$. The time span in which this branch is run through is about some μs. Concomitant, the nearly spherical fireball emerges attached to the anode. Since its emergence takes place under the aforementioned circumstances, i.e., for constant voltage of $\mathbf{U_3}$, the potential V of the anode marked **c** reveals the presence of a bifurcation node. This means that an infinitesimal variation around this value of V, produced by random phenomena (deterministic chaos always present in the plasma diode), decides the emergence or not emergence of the nearly spherical fireball. So, fixing the potential V at a value a little smaller than that marked **c,** the emergence of the nearly spherical fireball can be initiated by producing a minuscule spark between **E** and **A**. Once appeared, the nearly spherical fireball emphasizes certain robustness. This means that it survives when the voltage of $\mathbf{U_3}$ is varied, in certain limits, in both directions. So, the nearly spherical fireball survives when the voltage of $\mathbf{U_3}$ is decreased so that the branch marked **d-j** is run through. This phenomenon known as hysteresis proves that the nearly spherical fireball possesses a special kind of memory that enables it to perform all operations "learned" during its emergence by self-organization.

Returning to the premises under which the nearly spherical fireball emerges in the plasma diode, we remark that the first of them is the presence of an electric field sustained by $\mathbf{U_3}$

whose intensity must surpass a critical value. So, the plasma carries current owing to the work done by the dc power source $\mathbf{U_3}$. Under these circumstances, the population of electrons there present is driven away from thermal equilibrium so that the thermal energy distribution of the electrons becomes a quasi-maxwellian one. Accelerated in the electric field of the "point" anode (tungsten wire) the electrons achieve energies sufficient to produce ionizations and implicitly excitations of the atoms. As a result, a nearly spherical bipotential structure emerges at the distance from the anode where the ionization and excitation cross-sections functions of the electrons suddenly increase [33]. The bipotential structure emerged after accumulation of positive ions and electrons in adjacent regions sustains a proper drop of potential. Concomitant with its appearance, the symmetry related to the initially uniform spatial distribution of the plasma particles is broken, i.e., it transits into a radial one. The potential drop sustained by this bipotential structure grows concomitant with the voltage of $\mathbf{U_3}$ owing to the dependence of the ionization and implicitly the excitation rates on the anode voltage. However, because the increase of this potential drop requires collection by the anode of the electrons that produced and result after ionizations, the potential drop of the bipotential structure can increase only up the moment when the potential at the positive side equals the potential of the anode. In that moment, corresponding to the anode potential marked by **c** in Fig 5, the slope of the $I(V)$-characteristic abruptly changes. This value of V, that corresponds to a critical state of the plasma in the diode, initiates the appearance of a bifurcation-like instability after which the branch **c-d** is very quickly run through under constant voltage of $\mathbf{U_3}$. Simultaneously, the nearly spherical fireball appears attached to the anode. The investigation of its structure performed by using electrical probes revealed the followings: (i) A nearly spherical bipotential structure borders a nucleus enriched in positive ions; (ii) The bipotential structure sustains a potential drop in the order of the ionization potential of the argon gas; (iii) The potential at the positive side of the nearly spherical bipotential structure is several volts greater than the potential of the anode. (iv) The temperature of the electrons in the nucleus of the nearly spherical fireball is three times greater than the temperature of the electrons in the surrounding plasma; (v) The fireball is cold, i.e., the emission of light from its border is not caused by thermal excitation of the atoms.

Emerged in an environment driven at a critical distance from thermal equilibrium after a succession of events as non-equilibrium, instability, symmetry-breaking, bifurcation and long-range order, all of them considered key processes of self-organization [29], the nearly spherical fireball satisfies all requirements to be considered as a typical example of a self-organized complexity.

The nearly spherical fireball remains in a steady (ordered in behaviors) state when the voltage of $\mathbf{U_3}$ gradually increases up to the value of V marked by **e** in Fig 5. In this state, luminous sheaths, i.e., nearly spherical bipotential structures periodically peel off and re-form at the border of the nearly spherical fireball. Reaching the value **e,** the current I abruptly decreases (branch **e-f**) and, simultaneously, the fireball transits into a pulsatory state. Decreasing the voltage of $\mathbf{U_3}$ (branch **f-h**), the pulsatory state of the fireball persists till the potential V reaches the value marked **h** when, after an abrupt increasing of I (branch **h-i**), it returns to steady state. In the steady and pulsatory states, fireball generates coherent electromagnetic waves.

Subjecting the plasma in the diode (in a state before the nearly spherical fireball emerges) to the high frequency electric field created by the oscillator between the plates $\mathbf{E_1}$ and $\mathbf{E_2}$, the phenomena observed in the diode are different. Thus, gradually decreasing the power of the oscillator there is, as we will show later, a critical value for which a spherical flaming globe spontaneously emerges. Known as plasmoid [24, 25], the flaming globe floats freely between the electrodes $\mathbf{E}_1$ and $\mathbf{E}_2$. Its photograph is shown in Fig 6. The axial and radial potential distribution measured with movable emissive electric probes in the interior and the surroundings is shown in Fig 7.

A very important property of the plasmoid is its emergence and survival by a mechanism that involves resonant absorption of energy from the oscillator that generates the high frequency electric field. For illustrating this, we presented in Fig 8 a special kind of static characteristic [26] that emphasizes the change of power absorbed from the oscillator (measured by the current of the control grid I_C of the electronic tube that sustains the oscillations) as a function of the power of the oscillator (measured by the screen grid voltage V_S). This characteristic shows that the spontaneous emergence of the plasmoid can be revealed decreasing gradually (branch **a-b**) the power of the oscillator. When the critical value **b** is reached, the power absorbed from the oscillator increases abruptly (branch **b-c**) concomitant with the spontaneous emergence of the plasmoid. So, we prove that the plasmoid behaves as a spherical cavity able to absorb at resonance energy from the high frequency electric field created by the oscillator. The results of electrical probe measurements shown in Fig 7 reveal the presence of an enclosed bipotential structure that borders the plasmoid. It confers to the plasmoid the quality of a cavity able to absorb at resonance energy form the surrounding high frequency electric field created by the oscillator. An important fact revealed by these measurements is the ability of the enclosed bipotential structure that borders the plasmaoid to maintain the potential at its positive side greater than the potential inside its core. Another important fact is related to the abrupt increase of the power absorbed from the oscillator (branch **d-e**) after which the plasmoid transits to a new kind of dynamical state. In this state, spherical bipotential structures periodically peel off and re-form at the border of the plasmoid. They disintegrate after running through a certain distance from the border of the plasmoid. In this state the high frequency energy delivered by the oscillator is modulated with a frequency much smaller (in the order of tens of KHz) than that of the oscillator that surpasses one hundred MHz [26]. In this pulsating state, the plasmoid emits "signals" (electrical and optical ones) in the surroundings. These signals that can be receipted by other plasmoids at resonance contain information concerning the internal structure and behaviour of the plasmoid. When increasing V_S the value marked **f** is reached, the plasmid disintegrates concomitant with the abrupt decrease of the power absorbed from the oscillator (branch **f-g**). It is important to mention that the emergence of the plasmid can be initiated in the presence of a high frequency electric field also by an electric spark produced between **E** and **A**. In that case the nearly spherical fireball initially attached to **A**, detaches from it and transits to a free-floating plasmoid located between $\mathbf{E}_1$ and $\mathbf{E}_2$.

For initiating the emergence of the plasmoid we gradually decreased the energy of the high frequency oscillator till a state for which the additional plasma created by the high frequency electric field disappears. To avoid this critical state, a complexity (the plasmoid) emerges by self-organization. Bordered by a spherical bipotential structure able to accelerate electrons at energies sufficient to produce ionizations and implicitly excitations, the conditions for maintaining the additional high frequency plasma exist.

Considering the above described experimental results able to offer information concerning the spontaneous emergence and properties of the ball lightning, we will explain in the following the physical phenomena at their origin. For this reason we have in view the information offered by our previously performed experimental investigations [33-38] which have established that fireballs emerge in a plasma under the following circumstances (i) The population of the electrons in the plasma diode is a quasi-maxwellian one; (ii) The excitation and ionization threshold energies consist, besides the ordered kinetic energy of the electrons, also of their initial thermal energy; (iii) Concomitant with atom ionizations, the electrons increase the temperature of the population of electrons in the region where ionizations take place.

Taking into account these experimental finding it results that the electrons are able to transport the thermal energy achieved in the plasma and, using a part of it, to increase the temperature of electrons concomitant with the increase of the ionization rate. Evidently, because of their thermal motion, the ionization rate is increased only by that part of electrons momentum of which contributes to the increase of the energy of the electron producing the ionization process. Since the temperature of the electrons in the region where ionization takes place increases concomitant with the increase of the ionization rate, a plasma enriched in positive ions appears at that distance from **A** where the ionization cross section suddenly increases. This increase of the concentration of positive ions with respect to the concentration of electrons, concomitant with the increase of the ionization rate, takes place owing to: (i) The different mobility of the positive ions and electrons; (ii) The different thermal diffusivity of the positive ions and electrons. So, a group of positive ions appear in this region by a mechanism related to the local increasing of electrons temperature related to the ability of electrons to transport their thermal energy achieved in the plasma. Located in the next vicinity of the region where the electrons accumulated after atom excitation perform their thermal motion, there is the chance that a part of these electrons to electrostatic couple to positive ions. A great number of elementary electric dipoles self-assemble in this way [37, 38]. Every of them sequester a certain infinitesimal amount of electric field energy. Since the positive ions electrostatic coupled to electrons originate from thermal energy extracted by electrons from the surrounding plasma, the electric field energy sequestered in every dipole has its genuine origin in direct conversion of thermal energy into electric field energy. Acting in a concerted way, the dipoles initiate the formation of a nearly spherical bipotential structure in front of the anode. The potential drop supported by this bi-potential structure increases concomitant with the ionization rate, i.e., concomitant with the gradual increase of the voltage of $\mathbf{U}_3$ as long as the anode collects the electrons that produced and resulted after ionizations. This kind of gradual evolution of the nearly spherical bipotential structure is abruptly interrupted when the voltage of $\mathbf{U}_3$ reaches the critical value for which, through instability (abrupt increasing of I), the nearly spherical fireball spontaneously emerges.

Once emerged, the potential at the positive side of the bipotential structure at its border becomes greater than the potential of **A**. We explained this phenomenon [33, 37, 38] considering at its origin a self-enhancement of the number of positive ions and implicitly of the number of elementary dipoles. This takes place by a mechanism related to the presence in the surrounding plasma of a kind of electrons whose number increases quicker than the thermal energy of the electrons decreases. So, by successive acceleration of groups of this kind of electrons, a positive feedback mechanism is initiated. By this mechanism, the thermal energy extracted by groups of electrons from the plasma acts as the "catalyst" for increasing the number of positive ions located at the positive side of the bipotential structure. For working in this way, the slopes in all points of the descendant branch of electron thermal energy distribution function must be greater than 45^0.

Other experimental results potentially elucidative for explaining the appearance of the ball lightning are those obtained in the so-called low-voltage arcs [39]. Usually, in such devices the plasma was created by thermal ionizations at the surface of the heated cathode of a gas with small ionization potential, for example, cesium. The static $I(V)$-characteristic of such a diode is shown in the left side of Fig. 9. This characteristic reveals the presence of branches that abruptly change their slope. Performing photographs of the light phenomena observed in front of the anode and inside of the diode, images as that presented in the right side of Fig 9 were obtained. The numbers indicate the relationship between the observed light phenomena and different points of the static $I(V)$-characteristic. The importance of these kinds of plasma experiments resides in the fact that, after a succession of sequences similar to that in the plasma diode, a fireball that freely floats between the electrodes emerges by self-organization. The fireball transits into a pulsatory state, similar to that emphasized by the plasmoid, when the injected power is further increased.

Another kind of experiment, potentially informative for the emergence and properties of the ball lightning, we will present in the following. Thus, using the experimental device described in [40], schematically shown in Fig.10, it is possible to produce a well-located hot plasma in non-equilibrium in a metal vacuum chamber (pressure 10^{-8} Torr). In this experiment a Nd:Yag 10 ns pulsed laser beam has been focused on an Aluminum target producing there a well localized hot plasma. A cylindrical electric probe negatively biased with respect to the grounded vacuum chamber was connected to an oscilloscope. The formation and dynamics of the hot plasma in nonequilibrium created by the laser pulse was studied by using a fast ICCD camera placed perpendicular to its expansion. The ICCD camera was gated 20 ns at different delay with respect to the laser pulse. When the energy of the laser pulse exceeds a certain value (30-40 mJ) the results presented in Fig 11 were obtained. These results reveal that a well-located hot plasma in nonequilibrium is created by the laser pulse after local vaporization of the Aluminum target. Once emerged, the well-located hot plasma detaches from the surface of the target forming a nucleus surrounded by a plasma mantle. Afterwards that the nucleus evolves transiting through a succession of sequences into a free-floating fireball that survives a time span substantially longer than the time span of its generation by the laser pulse. Additional information concerning the actual origin of fireballs created by laser ablation techniques was obtained by using a cooper target [41]. The obtained results, presented in Fig. 12a,b, show that, beginning from a certain moment of time (700 ns) the fireball splits into two parts. Plotting the variation of the ionic current collected by the probe for different energies of the laser pulse the results shown in Fig 12b were obtained. These results prove that beginning from a critical value of the energy of the laser pulse (45 mJ) the ionic current collected by the probe reveals periodic variations. These variations are evidently related to the presence of moving nearly spherical bipotential structures that, during a limited time span, periodically peel-off from the border of the fireball. The aforementioned experimental results reveal strong similarities with that of the fireball emerged by self-organization in the plasma diode. Thus, in the plasma diode the fireball emerges by self-organization in a stable state when the injected energy reaches a first critical value. After additional injection of energy in the diode, the fireball transits into a steady state in which nearly spherical bipotential structures are periodic peeling off from its border. Starting from these similarities, the above presented laser ablation experiments become explainable, in our opinion, by considering at their origin a scenario of self-organization as that revealed by plasma experiments. Similar plasma spheres considered synthetic lightning balls, that persist in air up to a half second after the power source was turned off, are obtained in electric arc [42].

A very interesting experimental result proving the emergence of fireballs under conditions very different from these presented up yet was obtained at the Heavy Ion Collider in New-York, US [43]. There, beams of gold nuclei were smashed together at near light speeds. So the nuclei are broken into particles called quarks and gluons. These particles form a ball of plasma about 300 times hotter than the surface of the Sun. The fireball (Fig13) survives a time span that surpasses the duration of its emergence. This survival time of the fireball can be detected because it absorbs jet of particles produced by the beam collision. The model proposed to explain this experimental result considers that the particle are disappearing into the core of the fireball and reappearing as thermal radiation, just as matter is thought to fall into a black hole and come out as "Hawking" radiation.

The above briefly presented experimental results reveal that self-organization can appear under normal environmental conditions when: (i) A local cause, usually an electric spark, creates a well-located plasma by very quick injection of energy; (ii) The concentration and temperature of the well-located plasma must surpass those critical values for which a nucleus bordered by a plasma mantle emerges by a mechanism driven only by thermal energy; (iii) The electric field created by the nucleus must be so strong that the thermal distribution population of electrons has a descendant branch whose slopes in all point surpass 135^0. Under these circumstances, a self-organized complexity emerges through a bifurcation-like instability in a dynamical state. Once emerged in this way the dynamical state continues to be activated by thermal energy extracted by electrons from the plasma mantle. In plasma devices the plasma mantle is continuous supplied with thermal energy produced by work done by the plasma source so that the dynamical state of the self-organized complexity, i.e., its lifetime depends only on the laboratory conditions. Emerged in the same way under natural conditions the lifetime of the self-organized complexity is temporally limited by the amount of thermal energy initially located in the plasma mantle.

As revealed by the presented experimental results, the self-organization is a process the initiation of which requires the production of plasma with initial energy density and temperature that must surpass a critical value. The development of the self-organization process, once initiated, involves collective effects of quantum processes as ionizations and excitations. Consequently, the sizes of the emerged complexity depend, besides the amount of energy initially injected in the hot plasma in non-equilibrium, also on the environmental conditions where these quantum processes take place. So, under laboratory conditions but also in the Nature the sizes of the complexity emerged by self-organization could be very different, ranging from mesoscopic to macroscopic scales.

CONCEPTUAL MODEL OF THE BALL LIGHTING

In the following we relate the above briefly presented informational content offered by already published papers, including our published or submitted for publication ones, to the information offered by the very rare photographs of the ball lightning. Such a photograph presented in Fig 14 was taken by Sankt-Gallenkisch in Vorarlberg, Austria, in 1978 [11]. It shows that well-located hot plasma, revealed by a whitish center with a blue surround and a luminous tail, is produced in the point where linear lightning strikes the surface of the Earth. This photograph, that represents objective information concerning the emergence of the ball lightning, proves that the premise for its emergence is well located hot plasma (the whitish center) surrounded by a colder plasma mantle (the luminous tail). Produced by very quick injection of energy, the hot well-located plasma in non-equilibrium created by the linear lightning evolves naturally, i.e., in absence of an external driving force, into a direction that depends on its initial temperature and the time span in which the energy was injected. So, owing to the difference in thermal diffusivity of the electrons and the positive ions, the electrons leave the whitish center. As a result, a hot plasma enriched in positive ions surrounded by colder plasma enriched in electrons appears. When the initial temperature of the whitish center surpasses a critical value its potential becomes greater than the ionization potential of the surrounding gas. Implicitly, the electric field created by the whitish center is so strong that the thermal energy distribution of the electrons located in the plasma mantle becomes similar to that for which fireballs emerge by self-organization under controllable laboratory conditions [33, 37 38]. This means that the descendant branch of thermal energy distribution function of the electrons in the plasma mantle has slopes that surpass 135^0 in all points. Under such circumstances, a self-enhancement of the number of positive ions is initiated, by a mechanism similar to that present in a plasma diode, at that distance from the positive whitish center where the electrons extracted from the plasma mantle produce ionizations. Because concomitant with ionizations the electrons produce also a local heating, an important number of positive ions appear at the distance from the whitish center where the ionization cross-section function suddenly increases owing to the difference between the thermal diffusivity of electrons and positive ions. The positive ions are located in the next vicinity of electrons that have lost their kinetic energy by atom excitations and consequently are thermalized. So, a part of these electrons electrostatic couple, at a certain moment of their thermal motion, to the positive ions in excess located in the region where the ionization cross-section suddenly increases. In this way elementary electric dipoles are self-assembling, every of them sequestering a certain amount of electric field energy. Because the positive ions that form the dipoles appear after thermal diffusion of electrons, the sequestered electric field energy has its actual origin in a mechanism by which thermal energy extracted by electrons from the plasma mantle is directly converted into electric field energy [33, 37, 38]. Acting in a concerted way, the dipoles initiate the formation of a bipotential structure having its positive side located at the distance from the whitish center where the ionization-cross section suddenly increases and its negative side at the distance where the excitation cross-section function increases in the same way. So the whitish center is confined by an enclosed bipotential structure in a state of minimal free energy. Acting as a "membrane", the shape of the ball lightning is spherical one. Having its origin in a self-enhancement of positive ions, i.e., a positive feedback mechanism by which the number of dipoles abruptly increases, the spherical bipotential structure expands. This expansion takes place till the rate by which new dipoles appear equals the rate by which previously dipoles disintegrate after recombination of positive ions with electrons at the positive side of the bipotential structure. The survival of the ball lightning in this final phase of evolution is ensured by spherical bipotential structure that acting as a relative small membrane (cavity) ensures its stability based on a balance between: (i) The rate by which new dipoles are created by direct conversion into electric field energy of thermal energy extracted by electrons from the plasma mantle; (ii) The rate by which the previously self-assembled dipoles disrupt. The very small time span that exists between the moment when the dipoles self-assemble and the moment when they disintegrate by recombination with electrons located at their positive side ensures, in mean, the presence of the spherical bipotential structure at the border of the ball lightning.

Emerged by a mechanism as above described, the lifetime of the ball lightning depends on the "reserve" of thermal energy located in the plasma mantle. Consequently, depending on the initial conditions related to the production of the well-located plasma in nonequilibrium, the lifetime of the lightning balls could be very different.

Surviving in a dynamical state of self-organization, in which thermal energy is transported by electrons from a colder plasma (the mantle) to a warmer one (located at the positive side of the spherical bipotential structure), the "machinery" of the ball lightning works under circumstances in which the second law of thermodynamics does not work. In Nature, such conditions appear when plasma with density and temperatures surpassing a critical value is created locally by very quick injection of energy [37].

Considering the plasma mantle as a hot "bath" and the spherical bipotential structure as a '"giant" dipole the ball lightning reveals a kind of macroscopic quantum coherence similar to that stipulated by Fröhlich for explaining the origin of the most simple living systems [44].

Surviving in a dynamical state of self-organization, the ball lightning achieves macroscopic quantum coherence by a mechanism that involves: (i) Conversion with a constant rate of thermal energy, extracted by electrons from the plasma mantle, into electric field energy sequestered in dipoles; (ii) Emission with the same rate of electromagnetic energy after dipole disintegration by recombination of positive ions with electrons located at the positive side of the spherical bipotential structure. Formally, such a mechanism can be assimilated to that of a black hole that absorbs matter (electrons) concomitant with their thermal energy achieved in the plasma mantle and then radiates this thermal energy in the form of electromagnetic energy in the surroundings.

EXPLANATION OF BALL LIGHTNING CHARACTERISTICS

Considering the ball lightning as a self-organized complexity in a dynamical state, some (in our opinion all) of the observational characteristics listed in the literature become explainable. Thus:

- The very rare appearance of the ball lightning resides in the fact that the premises required for initiating its emergence by a scenario of self-organization, similar to that under which free floating fireball emerges in laboratory, are very strictly and consequently they can appear only under certain natural conditions.
- Its considerable stability is explainable considering the presence of a spherical bipotential structure the dynamical state of which is activated by a mechanism that involves extraction by electrons of thermal energy from the plasma mantle and its direct conversion into electric field energy. So, the life time of the ball lightning depends on the "reserve' of thermal energy initially located in the plasma mantle.
- The spherical shape of the ball lightning is determined by the presence of a bipotential structure in dynamical state characterized by a local minimum of the free energy

- Its dimensions are related to the gas pressure in the region where the linear lightning creates the well-located hot plasma in nonequilibrium. Since the ionization rate for which the expansion of the spherical bipotential structure at the border of the ball lightning stops depends on the gas pressure its sizes could be very different.
- Its color depends on the nature of the atoms located at the negative side of the spherical bipotential structure where the excitations are produced. Since the nature of the atom can vary, the color of the emitted light changes during the existence of the ball lightning.
- Its smell is related to ozone produced by ultraviolet light emitted concomitant with visible ones.
- The low temperature of certain lightning balls could be explained considering that, once emerged by self-organization, the collective quantum processes (ionizations and excitations at the two sides of the spherical bipotential structure) involved in sustaining its dynamical state are related to acceleration of electrons and only indirectly to their thermal energy.
- The frequently observed horizontal but occasionally erratic motion of the ball lightning could be explained considering that its existence is related to balance between: (i) The ascendant Arhimede forces present because of the rare gas located in the hot plasma created by the linear lightning; (ii) The attractive force that acts between the negative side of the enclosed bipotential structure and the positive Earth. Although present, the gravitational force is much smaller than the electrostatic one. Owing to the erratic causes related, for example, to the distribution of the Earth potential this balance is not the same.
- The extremely different energy densities associated with lightning balls could be related to the nature of the materials confined inside the spherical bipotential structure at their border. So, there are possible different nonlinear reactions initiated by the hot whitish center at the origin of the ball lightning. Confined in an enclosed bipotential structure, these reactions produced in dusty plasma, i.e., plasma that contains material particles with sizes in the range of nano/micrometers, could be very different. So, a burning process initiated in the whitish center at the origin of the ball lightning can prolong its lifetime.
- The occasionally observed motion of the ball lightning against the wind can be explained taking into account that the ionization rate sustained by electrons extracted from the plasma mantle depends on the concentration of the atoms. It is possible that the change of the concentration of the gas atoms at one side of the ball lightniung increases the ionization rate at this side. So, the permanent re-formation of the spherical bipotential structure could be different depending on the direction of the wind.
- The most astonishing property of the ball lightning, namely its capacity to penetrate windows without damage them could be explained considering that, once emerged by self-organization, the spherical bipotential structure at its border acts as a cavity able to absorb at resonance energy transported by electromagnetic waves, originating in external sources. Such waves are emitted, for example, by other lightning flashes produced simultaneously or immediately after the emergence of the ball lightning. Working as a 'cavity", the spherical bipotential structure locates by interference

energy in the form of strong electromagnetic oscillations. Reaching the surface of the window these "condensed" electromagnetic waves penetrates it. Consequently, a "knot" of electromagnetic waves appears at the other side of the window. It acts as the premise for the re-formation of the ball lightning at the other side of the window.

- The sound and radio interference occasionally associated with lightning balls becomes explainable considering its property to emit electromagnetic waves by transiting into a pulsating state similarly to fireball or plasmoid.
- The occasionally observed appearance of the ball lightning in higher regions of the Earth atmosphere can be potentially related to the electric activity in that region. Thus electrical discharge produced, for example, between clods cold easily create locally hot plasma in nonequilibrium capable to evolve by self-organization into a ball lightning
- The occasionally observed simultaneous appearance of linearly ordered lightning balls can be explained considering that the "channel" through which the linear lightning propagates is bordered by a cylindrical bipotential structure. Acting similar to surface tension, this cylindrical bipotential structure disintegrates in a set of "drops" under certain local conditions as for example sudden change of the speed or the direction of the wind.
- The very much energy release, occasionally observed during the disintegration of some lightning balls, attributed in the literature to nuclear reactions, could be explained by the capacity of the spherical bipotential structure to confine the matter of the whitish center whose temperature could be very high at densities for which such phenomena become possible.

CONCLUSIONS

Based on the informational content of laboratory experiments, we explain the appearance of the ball lightning considering at is origin a scenario of self-organization by which, through a bifurcation-like instability, a self-confined complexity emerges in a dynamical state. The capacity of the ball lightning to survive is related to the presence of the plasma mantle, simultaneously appearing with its emergence. Involving a mechanism by which electrons transport thermal energy from the plasma mantle to the warmer plasma located at the positive said of the spherical bipotential structure, the ball lightning survives as a self-organized complexity in a dynamical state performing operations not explainable by classical processes. In the vocabulary of thermodynamics, this means that the "machinery" of the ball lightning produces negative entropy. So, the assembly ball lightning and plasma mantle form a special kind of dissipative structure able to survive for a time span that substantially surpasses the duration of the self-organization process at its origin. Behaving as a cavity, the lifetime of the ball lightning can by prolonged by resonant absorption of energy.

Since well-located plasma in non-equilibrium can be produced in the atmosphere of the Earth, for example, by lighting flashes between clouds or electric sparks during earthquakes, lightning balls can appear during thunderstorms but also in absence of them. Other cause able to initiate the appearance of fireballs in the atmosphere of the Earth but also outside of it is the production of well-located plasma in nonequilibrium by collisions of objects that move with very high speeds in the space. When the temperature of the plasma created in this way surpasses a critical value, balls of fire could emerge by self-organization in a dynamical state whose lifetime depends on the amount of thermal energy located in the plasma mantle. So long lived fireballs can be occasionally observed in the atmosphere of the Earth. Since well-located hot plasma in non-equilibrium can be produced also in other parts of the Universe the appearance of so-called unidentified flying objects (UFO), becomes potentially explainable considering at its origin a scenario of self-organization similar to that at the origin of the lightning ball.

REFERENCES

[1] Roth J. Rece. *Ball lightning: what nature is trying to tell the plasma research community*, Fusion Technology 27, 255-270 (1995).

[2] Singer Stanley. *The nature of ball Lightning*". Plenum (1971)

[3] Barry, James Dale, *Ball lightning and Bead Lightning,* Plenum (1980)

[4] Stenhoff Mark. *Ball Lightning-An Unsolved Problem in Atmospheric Physics.* Kluwer Academic / Plenum Publishers (1999)

[5] Charman W. N. *Ball lightning*. Phys. Rep. (Rev. Sect. Phys. Lett.) 54, 261-306 (1979).

[6] Smirnov B. M. *The property and the nature of ball lightning*. Phys. Rep. (Rev. Sect. of Phys. Lett.) 152, 177-226 (1987).

[7] Turner D.J. *The structure and stability of ball lightning*. Phyl. Trans. R. Soc. London. A 347, 83-111 (1994).

[8] Sinkevich Oleg. *Long-lived plasma formation and problems associated with ball lightning*. High temperature, 35, 639-652, 955-969 (1997).

[9] Kadomtev B. *On the ball lighning phenomenon* Comments Plasma Phys. Control. Fusion. 13, 277-285 (1990).

[10] Lowke J.J. *Theory of ball lightning*. J. Phys. D Appl. Phys., 29, 1237-1244 (1996).

[11] Hubler G.K. *Fluff ball of fire*. Nature, 403, 487-478 (2000).

[12] Abrahamson I. and Dinniss J. *Ball lightning caused by oxidation of nanoparticle networks from normal lightning strikes on soil* Nature, 403, 519-521 (2000).

[13] Kapitsa P.L. *The nature of the ball lightning, Dokl. Akad. Nauk*. SSSR, 101, 245-248 (1955).

[14] Jennison , R.C., *Can ball lightning exist in a vacuum ?* Nature, 245, 95-96 (1973).

[15] Zheng X. H. *Quantitative analysis for ball lightning*. Phys. Lett.A, 148, 463-469 (1990).

[16] Ohtsuki Y. H. and Ofuroton H. *Plasma fireballs formed by microwave interference in air*. Nature, 383, 139-141 (1991).

[17] Yasui K. *Plasma fireballs feed by microwaves* Phys. Lett.A, 173, 451-455 (1993)

[18] Handel P.H. and Leitner J. F. *Development of the maser-caviton ball lightning theory*. J. Geophys Res. 99, 10,689-10,691(1994)

[19] Ranada A F. and Trueba J.I. *Ball lightning. An electromegnetic knot?* Nature, 383, 32-33 (1996)

[20] Altschuler M.D, House L.L. and Hildner, E. *Is ball lightning a nuclear phenomenon?* Nature, 228, 545-546 (1970).

[21] Torchigin V.P. and Torchigin A.V. *Physical nature of ball lightning*, Eur. Phys. J. D, 36, 319-327 (2005).
[22] Strattman W. *Irving Langmuir's ball lightning tube*, I Langmuir's Ball Lightning Tube -Microsoft Internet Explorer 22 05 1996
[23] Song B, D'Angelo N and Merlino R. L. *On anode spots, double layer and plasma contactors,* J Phys D: Appl. Phys. 23, 1789-1795 (1991)
[24] Lozneanu E, Popescu V, Popescu S and Sanduloviciu M. *Spatial and spatiotemporal patterns formed after self-organization in plasma*. IEEE Trans. Plasma Sci. 30, 30-31 (2002)
[25] Taillet J. *Resonance–sustained radio frequency discharges,* American J. Phys. 37, 423-431 (1969)
[26] Sanduloviciu M. *Pulsatory plasmoid formed in high frequency electric field,* Plasma Phys. Contr. Fusion 29, 1687-1694 (1987).
[27] Dikhtyar V. and Jerby E. *Fireball ejection from molten hot spots in air by localized microwaves*. Phys. Rev. Lett. 96, 045002-1. 045002-4 (2006)
[28] Reucroft S. and Swain J. *Microwaves create ball lightning*. CERN COURIER. 46, No 3, p 9 (2006).
[29] Nicolis G and Prigogine I. *Exploring Complexities* (W. H. Freeman and Co., New-York (1989)
[30] Heylighen Francis. *Complexity and Self-organization,* Edited by Marcia J. Bates and Mary Niles Maak. (Taylor & Francis, 2008)
[31] Juettner B, Noack C, Versteegh A and Fusssmann G. *Long-living plasmoids from a water discharge at atmospheric pressure.* Proc. 28th. Int. Conf. Phen. Ionied Gases, Prague, Cczech. Republic 2229- 2230 (2007)
[32] Sanduloviciu M and Lozneanu E. *The ball lightning as a self-organization phenomenon.* J. Geophys. Res. Atmos.105, 4719-4727 (2000).
[33] Lozneanu E and Sanduloviciu M. *Self-organization scenario acting as physical basis of intelligent complex systems in laboratory.* Chaos, Solitons and Fractals 30, 125-132 (2006).
[34] Sanduloviciu M and Lozneanu E. *On the generation mechanism and the instability properties of anode double layer.* Plasma Phys. Control. Fussion 28, 585-595 (1986)
[35] Lozneanu E, Popescu V, and Sanduloviciu M. *Negative differential resistance related to self-organization phenomena in a dc gas discharge.* J. Appl. Phys. 92, 1195-1199 (2002)
[36] Lozneanu E, Popescu S and Sanduloviciu M. *Ball lightning as a self-organized complexity.* Unifying Themes in Complex Systems. Vol IIIb. Eds, A. A Minai, Y. Bar-Yam, Springer Verlag, Berlin Heidelberg, USA, 2007, pp.129-137
[37] Lozneanu E, and Sanduloviciu M. *Self-organization scenario grounded on new experimental results.* Chaos, Solitons and Fractals. 40, 1845- 857 (2009)
[38] Lozneanu E and Sanduloviciu M. *Plasma experiments with relevance for the physical basis of electric memory and macroscopic quantum coherence.* Submitted for publication.
[39] Kiryuskchenco A. I. and Lebedev M. A. *Low voltage arc in cesium vapor I Formation of spherical plasmoid.* Soviet Phys. Techn. Phys. 16, 924-926 (1971).

[40] Ursu C, Gurlui S, Focsa C and Popa G. *Space-and time-resolved optical diagnosis for the study of laser ablation plasma dynamics* Nucl. Instr and Meth. in Phys.Res.B 267, 446-450 (2009)

[41] Gurlui S, Sanduloviciu M, Strat M, Mihesan C, Ziskind M, Focsa C. *Periodic phenomena in laser alblation plasma plums; A self-organization scenario.* AIP. Conf. Proc B12, 279-272 (2006)

[42] Alexeff I, Parameswaran S, Thiayagrajan M, Grace M. *An obsevation of synthetic ball lightning.* IEEE Trans. Plasma Sci. 33, 498-499 (2005)

[43] Nastase H. *Lab fireball may be black holes* http://news.bbcco.uk/1/hisci/tech/435761.stm

[44] Fröhlich H. *The biological effects of microwaves and related questions. Adv* Electronics and Electron Phys, 53, 85-152(1980)

In: Lightning: Properties, Formation and Types
Editor: Matthew D. Wood
ISBN: 978-1-61122-198-5

Chapter 2

EXTRATERRESTRIAL LIGHTNING AND ITS PAST AND FUTURE INVESTIGATION

Georg Fischer[1], Donald A. Gurnett[2] and Yoav Yair[3]
[1]Space Research Institute, Austrian Academy of Sciences, Graz, Austria
[2]Dept. of Physics and Astronomy, University of Iowa, Iowa City, USA
[3]Dept. of Life and Natural Sciences, Open University of Israel, Ra'anana, Israel

ABSTRACT

All extraterrestrial lightning detected so far on the planets Jupiter, Saturn, Uranus, Neptune, and potentially at Venus is attributed to intracloud lightning. On Mars other types of discharges might exist within dust storms. In this commentary we will focus on the historical development of planetary lightning investigations at Jupiter, Saturn, and Venus. Special attention will be devoted to the controversies around the existence of lightning on Venus. Thereby we will highlight the importance of combined optical and radio observations for the epistemological degree of confidence in lightning detection from space as well as for its detailed investigation. Future spacecraft missions to various planets should take this into account. The main future extraterrestrial lightning investigations will be outlined. We also plead for radio observations with a time resolution of the order of microseconds and for more modelling and laboratory work in this field.

1. INTRODUCTION

The most essential characteristic of a lightning flash is its optical emission that is caused by a large electrical current of an atmospheric discharge. The word "lightning" contains the word "light" and thereby emphasizes optical emissions, despite the fact that lightning discharges radiate in a broad spectrum from radio waves to gamma rays. This supremacy of optical emissions might simply be explained by the importance of the sense of sight for humans. The usefulness of radio emissions becomes immediately evident when one makes the nice experiment of turning on an AM radio receiver during a thunderstorm: Each time one sees a lightning flash there should be a crackling sound in the radio as well. But, not every time one hears the crackling sound is the lightning flash visible, because it can be hidden somewhere in the thundercloud at a far distance. Clearly, radio observations are better for counting lightning discharges, and that is why lightning detection networks usually rely on electric or magnetic antennas optimized for radio wave reception in the frequency range of a few kHz where the lightning channel current is an efficient radiator.

This paper is not intended to be a comprehensive review paper of planetary lightning since an updated review about this subject was recently published by Yair et al. [2008]. However, we will shortly narrate the history of lightning detection on the gas giants and Venus and highlight the controversies. In the next section we will briefly touch various types of extraterrestrial lightning where actually much is still unknown. Then we will show that lightning on Jupiter and Saturn have been detected both by optical and radio wave measurements, whereas for Uranus and Neptune only the latter is available. One section is devoted to the controversial case of Venus lightning, and another one will view the problem from a kind of epistemological point of view. This will lead us to multi-instrument and multi-phenomenology strategies for future investigations of extraterrestrial lightning which will be outlined in the last section.

2. Types of Extraterrestrial Electrical Discharges

2.1. Lightning

Lightning on the gaseous Jovian planets (Jupiter, Saturn, Uranus, and Neptune) necessarily must be of intracloud type (IC), since these planets do not have a solid surface. But also lightning on Venus should rather be of intracloud type. The high atmospheric pressure and the distance of the cloud layer around 45-70 km altitude to the ground would require an unrealistically high electrostatic field for a cloud-to-ground (CG) breakdown [Yair et al., 2008]. On Earth, cloud discharges are not as extensively investigated as cloud-to-ground discharges since they pose no threat for life and infrastructure except for airplanes. Studies relating HF (high frequency) radiation and optical emissions to processes during intracloud flashes like the one by Shao and Krehbiel [1996] or Light et al. [2001] could be important for comparisons of terrestrial with extraterrestrial intracloud lightning.

The only place so far where extraterrestrial cloud-to-ground lightning is considered to be possible is Saturn's enigmatic moon Titan [Tokano et al., 2001]. Fulchignoni et al. [2005] reported several impulsive electric field events detected by the Huygens Atmospheric Structure Instrument (HASI) during the descent of the Huygens Probe through Titan's atmosphere, but their origin is still unclear. A recent claim by Morente et al. [2008] of having identified Schumann resonances in the HASI data has been forcefully rejected by Hamelin et al. [2009] for the HASI team. The Saturn orbiting Cassini spacecraft is equipped with a camera and a radio instrument. The Imaging Science Subsystem (ISS) [Porco et al., 2004] and the Radio and Plasma Wave Science instrument (RPWS) [Gurnett et al., 2004] should be able to detect lightning on Titan. However, no optical or radio emissions indicative of lightning were recorded during any of the dozens close Titan flybys [Fischer et al., 2007a].

2.2. Transient Luminous Events

Another interesting area of research are TLEs (transient luminous events like sprites, jets, or elves caused by lightning) that have not yet been detected on any other planet. On Earth sprites are related to large positive cloud-to-ground flashes, and such types of flashes probably don't exist on other planets. However, carrot sprites have been shown to be also related to intracloud activity [Van der Velde et al., 2006], and slow intracloud discharges with large charge moment changes might as well fulfil the conditions for sprite initiation. First model calculations by Yair et al. [2009] have predicted the appearance of sprites in the upper atmosphere of Venus and Jupiter. Most recently, Dubrovin et al. [2010] conducted laboratory experiments on streamers in various gas mixtures to determine emission spectra for potential sprites at Venus and at the Jovian planets. Maybe the strong radio emissions of Saturn lightning can trigger elves [Fischer et al., 2008], a possibility which still needs further investigation. It is clear that research on extraterrestrial TLEs has just started and there might be still a long way to go until their actual detection in alien atmospheres.

2.3. Lightning in Dust and Volcanic Plumes

Other exotic extraterrestrial types of lightning in a wider sense would be volcanic lightning and discharges in dust storms. On Earth, lightning is ubiquitous in volcanic eruptions, with some spectacular features observed within the ash cloud [Thomas et al., 2007]. The idea that potential Venus lightning is linked to volcanic activity [Ksanfomaliti et al., 1983] was later rejected by Borucki [1982], Taylor et al. [1985], and Russell [1991]. According to Peratt and Dessler [1988] volcanic plumes on the Jovian satellite Io could be a visible manifestation of a plasma-arc discharge. Dust storms on Mars could exhibit electric discharges [Melnik and Parrot, 1998; Farrell et al., 1999a], and evidence for the emission of non-thermal microwave radiation by a Martian dust storm was recently published by Ruf et al. [2009]. This radiation might be related to electrical discharges in the dust storm, but a more direct confirmation is still missing. A four-year search for impulsive radio signals from lightning at Mars has been conducted by Gurnett et al. [2010] using the radar receiver on the Mars Express spacecraft. Mars Express is in orbit around Mars. The receiver, which is part of the Mars Advanced Radar for Subsurface and Ionosphere Sounding (MARSIS) has an integration time of 91 μs, a bandwidth of 10.2 kHz, and can detect radiation down to the galactic noise level. So far, after a total of more than 140 million measurements spanning all longitudes and latitudes, and altitudes from 275 to 1000 km, no lightning has been detected.

3. Lightning Detection on the Jovian Planets

3.1. Jupiter

The supremacy of optical lightning detection is clearly evident when looking at the case of Jupiter: No one seriously questioned the detection and interpretation of bright spots on Jupiter's night side by the Voyager 1 camera [Smith et al., 1979; Cook et al., 1979] as being due to lightning discharges in Jupiter's atmosphere. Additional confirmation came from the detection of whistlers [Gurnett et al., 1979], although most flashes recorded by the plasma wave analyzer were not seen by the Voyager camera [Borucki et al., 1982]. Subsequent spacecraft missions demonstrated the easiness of optical flash detection at Jupiter: The cameras of Voyager 2 [Borucki and Magalhães, 1992], Galileo [Little et al., 1999], Cassini [Dyudina et al., 2004], as well as New Horizons [Baines et al., 2007] spotted lightning on both hemispheres up to high latitudes around 80° with particular active regions near 50° in areas of cyclonic shear or near the centers of westward jets. The lightning and radio emission detector (LRD) of the Galileo probe observed lightning sferics in the frequency range of a few kHz inside Jupiter's atmosphere [Lanzerotti et al., 1996; Rinnert et al., 1998].

3.2. Saturn

The situation at Saturn was a bit more complicated. Radio signals attributed to Saturn lightning were first detected by the radio instrument onboard Voyager 1, and Warwick et al. [1981] termed the impulsive short radio bursts as Saturn Electrostatic Discharges (SEDs). Initially there were some doubts about their origin, and a source in Saturn's B ring was suggested by Evans et al. [1982]. But, Kaiser et al. [1983] showed with an argument of visibility that the Voyager SEDs should stem from an atmospheric source in Saturn's equatorial region. This explanation was largely accepted by most scientists, although no corresponding storm was imaged by the Voyager cameras. Since the radio waves traverse Saturn's ionosphere on their way to the spacecraft, Kaiser et al. [1984] determined the peak electron densities of Saturn's ionosphere as a function of local time from the measurement of the low-frequency cutoff below which the SEDs are not detected. The result was a diurnal variation in electron density of more than two orders of magnitude. Majeed and McConnell [1996] contended that the SEDs might not come from Saturn's atmosphere, because this large diurnal variation could not be explained by their ionospheric model due to the long chemical lifetime of the dominant H^+ ion relative to a 10-11 hour Saturn day. Similarly, Rakov and Uman [2003] noted that SEDs alone might not be a convincing evidence of Saturn lightning due to the lack of whistler and optical observations. This critique seems rather rigorous, but has a certain justification since nearly every space mission investigates uncharted territory. Strictly speaking one cannot completely rule out a different kind of radio source, although the SEDs showed similarities to terrestrial lightning. The SEDs usually appear in episodes lasting about half a Saturn rotation when the storm is on the side of the Saturn facing the spacecraft. Figure 1 shows such an SED episode in a so-called dynamic spectrum, where the color-coded intensity of the radio waves is displayed as a function of time and frequency. It was recorded by the Cassini RPWS instrument in August 2009, and the short vertical bursts above ~1.8

MHz are the SEDs. Intrinsically the SEDs are broad-banded bursts but they appear as narrow-banded due to their short duration and the frequency sweeping of the receiver.

In case of the SEDs, the Cassini mission could remove all the doubts about their origin. The Cassini camera imaged bright storm eruptions in Saturn's "storm alley" at a planetocentric latitude of 35° south that correlated with the occurrence of SEDs [Porco et al., 2005]. Combined observations of Cassini ISS and RPWS revealed consistent longitudinal drifts of storm clouds and SED episodes [Dyudina et al., 2007; Fischer et al., 2007b] plus a good correlation between the brightness of the storm clouds with the number of detected SEDs. For most SED storms ISS spotted a corresponding cloud feature, and no prominent clouds were observed during SED inactivity. Direct flashes of light were finally identified in night side images of an SED storm cloud around Saturn's equinox in August 2009 [Dyudina et al., 2010]. The reduced ring shine allowed this first visible detection of Saturn lightning, whose source is located deep in the atmosphere, some 125-250 km below the cloud tops around the freezing level of the water cloud. Figure 2 compares images of Jovian and Saturnian flashes taken on the night side of the respective planet. The size of the flash-illuminated cloud region is larger for Saturn than for Jupiter pointing to a deeper flash source for the former. The Saturn lightning flashes in Figure 2 were indeed detected during the SED episode that is displayed in Figure 1.

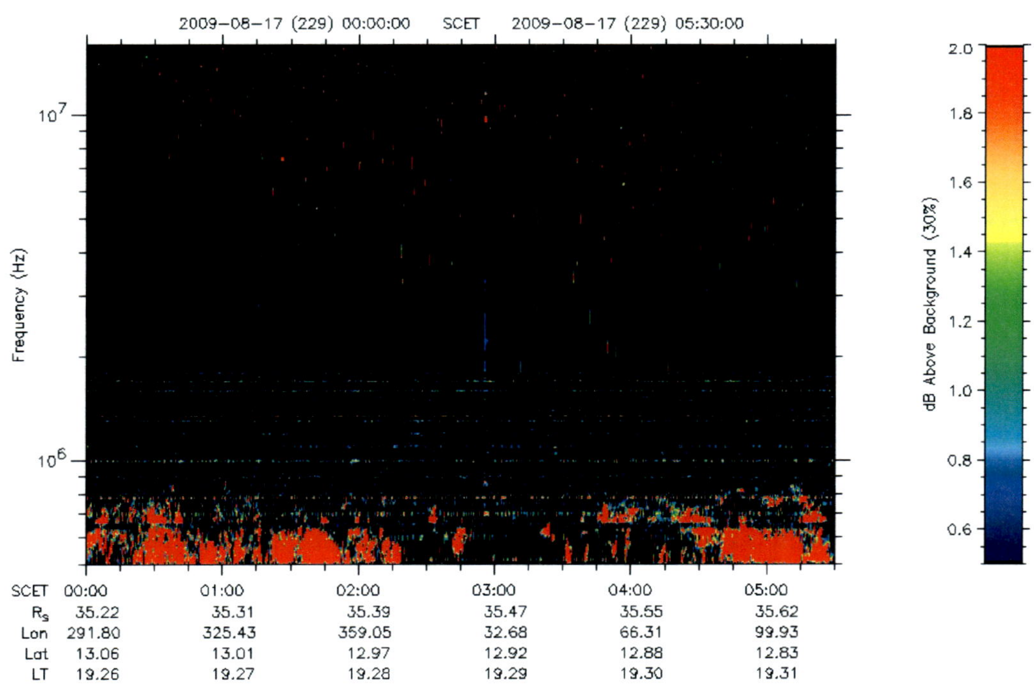

Figure 1. A dynamic spectrum of an SED episode recorded by Cassini RPWS on August 17, 2009. The color-coded intensity is plotted as a function of time and frequency. Besides the spacecraft event time the abscissa shows the orbital parameters of Cassini, which are the distance in Saturn radii (R_S), the sub-spacecraft western longitude, the latitude, and the local time

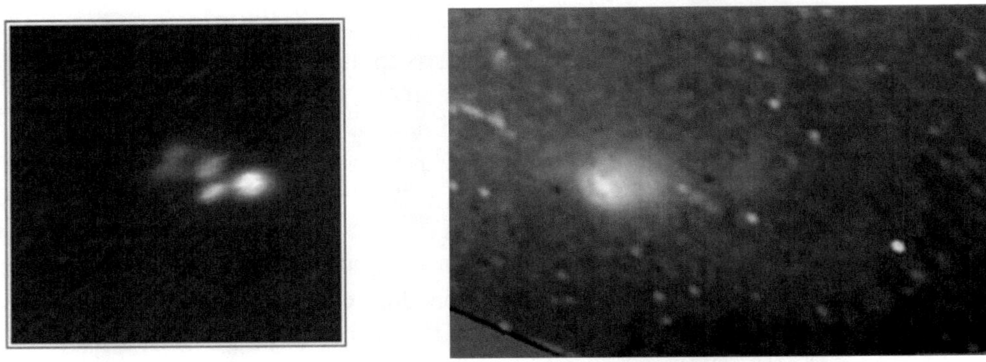

Figure 2. Comparison of optical flash detections on the night sides of Jupiter (left) and Saturn (right). The images were taken by the cameras onboard Galileo [taken from Little et al., 1999] for Jupiter and onboard Cassini [taken from Dyudina et al., 2010] for Saturn. The center of the Galileo image is located around 48° north and 17° west, and it spans a region of 2.4° in longitude (~1900 km) and latitude (~2900 km). The center of the Cassini image is at 36° south and 11° west, and it spans 3.5° in longitude (~2900 km) and 2.0° in latitude (~2000 km). The illuminated cloud regions have a size of a few hundred kilometers, whereas the very small spots are due to cosmic ray hits.

Interestingly, Jovian lightning seems to have no HF radio component like the SEDs are for Saturn. This could be either due to radio wave absorption in Jovian's ionosphere [Zarka, 1985] or Jovian lightning could be a slow discharge with marginal radiation in the HF part of the spectrum [Farrell et al., 1999b]. At Saturn only one whistler was detected [Akalin et al., 2006], and the scarcity of whistler observations is most likely a trajectory issue since the spacecraft has to pass through the storm-connecting magnetic field lines. The recent imaging of visible lightning at Saturn has brought the level of confidence in Saturn lightning detection to the same level as Jovian lightning detection. This is totally different for lightning on Venus, whose existence has now been debated for about three decades.

3.3. Uranus and Neptune

The confidence in lightning detection on Uranus and Neptune is similar as it was for Saturn before the new findings of Cassini. Only the Voyager 2 spacecraft visited the two distant giants Uranus and Neptune. At Uranus about 140 HF signals similar to SEDs were detected and termed UEDs for Uranus Electrostatic Discharges [Zarka and Pedersen, 1986]. At Neptune the indications for lightning are due to radio emissions as well, and 16 whistlers [Gurnett et al., 1990] and 4 potential HF signals [Kaiser et al., 1991] were detected.

4. The Controversial Case of Venus Lightning

Venus has a thick cloud cover that encircles the whole planet and is the most likely place for lightning to occur. The cloud deck is mainly composed of droplets of sulphuric acid (H_2SO_4) which has a high dielectric constant around 100 (in liquid form) due to the high polarizability of the molecule. There are three main cloud layers extending from 70 down to

45 km in altitude at pressures from 20 to 1000 mbar and temperatures from 220 to 400 K. There is a strong zonal flow and the cloud deck super-rotates and travels around the equator within 4-5 Earth days. The clouds are of stratiform morphology, and there is little evidence of potential latent instability [Esposito et al., 1983] as it is present in terrestrial thunderclouds. Most of the visible radiation is reflected from the upper and middle cloud layer, and at deeper atmospheric levels particles with radii around 1 μm contribute most to the optical thickness leading to a total optical depth of ~30 at the surface [Hueso et al., 2008]. Recently, Michael et al. [2009] investigated the electrification of cloud particles at Venus. Their model led to more negatively charged particles in the upper cloud layer compared to below, but the electric field would be too low for a breakdown. It has been often noted that the weakness of vertical convection would not allow a large-scale separation of charged particles unfavourable for the generation of lightning discharges. Small convective cells with a horizontal extent of about 20 km were detected in the upper cloud layer by the Venus Monitoring Camera onboard Venus Express [Markiewicz et al., 2007].

4.1. Electromagnetic Detection of Lightning-Related Signals

4.1.1. The Venera missions

Indications of lightning on Venus come from tentative optical detection plus three different kinds of radio waves, which are sferics of a few kHz detected by atmospheric probes, whistlers detected by orbiting spacecraft, and HF radiation that was only measured by spacecraft flying by. It all started with impulsive signals measured by the magnetic loop antenna of the Groza instrument in the frequency range from 8-100 kHz by the Soviet Venera 11 and 12 descent probes [Ksanfomaliti et al., 1979; Ksanfomaliti, 1980]. The lightning interpretation is questioned by the fact that signals with a rate of ~20 s^{-1} were detected during descent, but nearly none after landing. The suggested inhibition of direct propagation by the horizon is considered unlikely, since only special geometries and beaming would lead to ionospheric reflections not being detected also beyond the horizon. The cessation of the signals after landing (not only for Venera 11 and 12, but also for Venera 13 and 14) leaves the strong suspicion that the signals attributed to lightning sources were actually generated by the descent of the probes themselves [Hunten, 1995]. In the paper where the Venera 13 and 14 results are presented, Ksanfomaliti et al. [1982] do not consider lightning as the reason for the low frequency noise, and they simply call it an electrical-discharge source being located beneath the cloud deck.

4.1.2. Pioneer Venus Orbiter (PVO)

The detection of signals in the 100 Hz channel of the plasma wave instrument onboard the Pioneer Venus Orbiter (PVO) was attributed to lightning whistlers by Taylor et al. [1979] and Scarf et al. [1980]. The controversy about the origin of these signals was sparked by H.A. Taylor et al. [1985] who pointed out the strong association of the 100 Hz bursts with night side ionospheric troughs. They suggested that the source is not lightning but a local plasma instability associated with the steep density gradient at the edge of the trough. Scarf [1986] commented on that paper and Taylor et al. [1986] answered with a reply. Many subsequent papers were published [e.g. Singh and Russell, 1986; Taylor and Cloutier, 1986; Taylor et al.

1987; Scarf and Russell, 1988; Taylor and Cloutier, 1988; Russell et al., 1989; Russell, 1991; Taylor and Cloutier, 1992] from both sides, but the issue has not been satisfactorily resolved [Hunten, 1995]. In his reviews Russell [1991; 1993] even considers the bursts at 730 Hz as due to lightning, because in this way the non-detection with the PVO star sensor [Borucki, 1982] and the VEGA balloons [Sagdeev et al., 1986] in the post-midnight to dawn region could be explained since Venus lightning should be preferentially around dusk. But, 730 Hz should be above the electron cyclotron frequency where propagation in the whistler mode is not possible, and it remains dubious how the 730 Hz bursts could propagate through the ionosphere to the spacecraft. The PVO plasma wave instrument did not provide waveforms, and therefore could not confirm the existence of dispersion, a fundamental characteristic of whistlers.

If the signals detected by the Pioneer Venus Orbiter are not due to lightning whistlers, what else could they be? Maeda and Grebowski [1989] provided an answer and presented whistler-mode emissions called "saucers" that are observed in the terrestrial polar magnetosphere. These signals are located in night-side troughs where the magnetic field is radial [Gurnett and Frank, 1972], very similar to the Venusian 100 Hz emission. Maeda and Grebowski [1989] showed that for a receiver with a fixed frequency channel with a narrow bandwidth the typically V-shaped saucers would turn into individual bursts of a fraction of a second long, just like the 100 Hz signals that were observed at Venus. Another possibility was offered by Huba [1992] who considered a lower-hybrid-drift instability as a viable mechanism. These two possibilities were refuted by Strangeway [1992] on theoretical grounds.

4.1.3. Venus Express

Similar bursts were detected again by the magnetometer onboard Venus Express and Russell et al. [2007] attributed them to lightning whistlers. Their duration is 0.25-0.5 s, and they are right-hand polarized consistent with whistler-mode waves. The access of the waves to the night side ionosphere of Venus is controlled by the angle of the local magnetic field with respect to the horizontal, which has to be greater than 15° for the bursts to appear [Russell et al., 2008]. So far these new measurements have not resulted in as much controversy as before, because Venus Express has no plasma wave package and also cannot measure plasma densities that could identify the ion troughs. However, Yair et al. [2008] contended that the duration of the whistler signals is much shorter than terrestrial lightning whistlers measured by rockets at comparable altitudes of a few hundred kilometres by Holzworth et al. [1999].

4.1.4. The Galileo and Cassini fly-bys

HF signals that could possibly be attributed to lightning discharges were detected by the radio instrument of the Galileo spacecraft during its short flyby [Gurnett et al., 1991]. Galileo only identified a handful of weak bursts from a distance of 4 to 5 Venus radii, and it was not possible to see if the burst's intensity approximately decreased with distance squared as it would be required for a signal originating from Venus. The two flybys of Cassini at Venus did not result in any unambiguous burst detection [Gurnett et al., 2001]. The Venus 1 flyby of Cassini in April 1998 with an altitude of 300 km was particularly close and mainly at the night side, i.e. nearly in an optimum position for lightning detection, but the result was negative. Figure 3 shows the electric field intensities measured by RPWS during this flyby for 20 frequency channels. The various types of radio emissions are indicated in the plot. The only potential Venus lightning burst measured at 2.225 MHz around 13:46 UT is also plotted in a separate box. However, such a single unidentified burst should not be taken as a proof for lightning. Note that this single burst was detected when Cassini was just 400 km above the surface of Venus, while Figure 1 shows several Saturn lightning bursts measured by the same instrument from a distance of more than 2 million kilometres. This should illustrate the strength of Saturn lightning as well as the sensitivity of the radio instrument. The same Cassini instrument also detected several hundred HF radio signals from terrestrial lightning within 14 Earth radii during the close Earth flyby in August 1999. So far, there has not been a Venus orbiting spacecraft equipped with a radio instrument capable of detecting HF bursts.

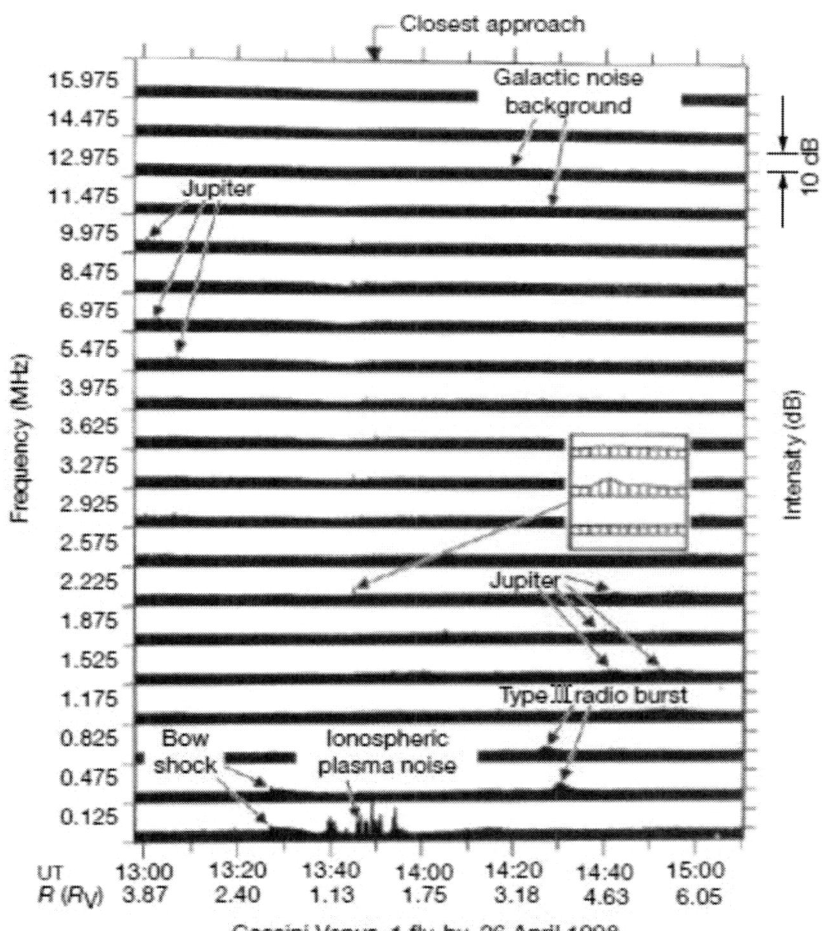

Figure 3. A 20-channel plot from Gurnett et al. [2001] of the Cassini RPWS electric field intensities (in dB) as a function of time (in UT) during the close Venus-1 flyby. The frequencies are indicated on the left side, and at the abscissa the distance to the center of Venus in Venus radii (R_V) is given below the time in UT. Various types of emissions that should not be mixed up with lightning are indicated

4.2. Detection of Optical Signals

The first tentative optical lightning detection was made by the airglow spectrometer on Venera 9 that detected several pulses during a 70 s period [Krasnopolsky, 1983]. The effect was not present when the instrument was pointed away from Venus. The characteristic duration of signals was ~0.25 s, but there was no typical substructure like for terrestrial lightning although the instrument had a sampling interval of 20 ms. Other optical instruments failed in lightning detection: The PVO star sensor [Borucki, 1982] detected events related to energetic particles from solar flare events and none could be attributed to lightning. The two VEGA balloons [Sagdeev et al., 1986] did not detect any optical flashes from lightning during their cruise through the middle cloud layer at an altitude of 54 km that comprised 7 hours of observations. No flashes were detected by the camera onboard Galileo during its flyby in 1990 [Belton et al., 1991]. So far the camera onboard Venus Express has not detected lightning [Titov, personal communication] although it thoroughly investigates the Venusian atmosphere. Terrestrial intracloud lightning illuminates its surrounding clouds, and sometimes the discharge path is even outside of the cloud. There could be also cloud-to-air lightning which would facilitate optical detection. Taking a similar rate of lightning of Venus with respect to Earth as suggested by various authors, it is strange that an orbiting spacecraft has not yet made an unambiguous optical detection of lightning on the night side of Venus. Williams et al. [1982] studied the optical properties of Venusian clouds by a Monte Carlo simulation. They found that the non-detection of optical signals from space is not due to attenuation by the atmosphere and the clouds, and that lightning is not an acceptable explanation for Ashen light. The second tentative optical Venus lightning detection was made by Hansell et al. [1995] by telescopic observation from Earth, which is an experimentally complicated procedure due to scattered light from the day side. Six tentative flashes were detected with optical energies of 10^8 to $2 \cdot 10^9$ J from the Venus cloud tops. We note that the optical energy is comparable to that of flashes detected at the Saturnian cloud tops by Dyudina et al. [2010], i.e. Venus orbiting spacecraft equipped with cameras should be able to detect flashes with such optical energies as well.

5. An Epistomological Point of View on the Problem

Russell [1993] notes that the controversy about the existence of Venus lightning is surprising since evidence can be found in all three sets of optical, VLF, and radio emission data. However, all observations are single observations that in each case could have an alternative explanation. No combined observations yet exist, where optical and radio emissions of potential Venus lightning were detected at the same time and could be linked to each other by temporal or spatial correlation. These combined observations are present for Jovian and Saturnian lightning. Hence, it should not be surprising that the degree of confidence in Venus lightning detection is clearly lower compared to Earth and these two giant planets. For all those detections with single instruments at Venus where other possibilities could not be ruled out, a parallel detection with a second instrument would have convinced the sceptics. Scepticism is a healthy thing in science, and for the sceptic it is impossible to prove the non-existence of a certain phenomenon. But, the burden of proof is not on the side of the sceptic.

A non-contradictory picture of lightning on Venus has yet to emerge. E.g. the optical observation of 6 flashes within 3 hours from the major part of the night side of Venus by Hansell et al. [1995] is in stark contrast with the data from the Venera probes where Ksanfomaliti et al. [1979] suggested 20 flashes per second originating from a single storm front. Despite the huge number of publications on potential Venus lightning we still know relatively little about it. It is not known if Venus lightning is a permanent phenomenon or if there are times when no thunderstorms are present. The various non-detections suggest that the latter is probably the case. This might explain the non-detection by Cassini, but another possibility would be that Venus lightning has no HF component which would imply that Galileo detected spurious HF bursts not due to lightning. The limited coverage of each potential Venus lightning observation makes all estimations of global flash rates very hypothetical. The presence of lightning on the day side of Venus is unknown since optical and whistler observations can only be made at the night side, and the potential sferics observed by the Venera probes on the day side could be the result of propagation from the night side to the day side by ionospheric reflections. It is worth noting that convective cells as likely places for lightning are rather typical for low latitudes just downstream from the subsolar point [Esposito et al., 1983], which is in contrast to the claimed preference of lightning on the dusk to night side by Russell [1991].

The duration of the PVO whistler-mode signals and the duration of the optical signals detected by Venera 9 (~0.25 s) without a substructure suggest that Venus lightning might be somewhat different to Earth lightning, and it could be a slower discharge. It is reasonable to be somewhat flexible in this respect and not to only expect physical properties similar to Earth lightning. A little bit more flexibility and openness to other possibilities should be considered as well when talking about the origin of the whistler-mode signals. Given the fact that we still know very little about the Venus night side ionosphere, it should not be sufficient to say that we know of no other processes other than lightning that could cause such whistler-mode signals. In both cases, physical properties of Venus lightning as well as origin of the whister-mode signals, we should try to go somewhat beyond the saying "you only see what you know" (attributed to Johann Wolfgang von Goethe).

In Table 1 we summarize the potential detections of lightning on the various heavenly bodies discussed in this paper. Combined observations of optical and radio data has been only made for lightning on Earth, Jupiter, and Saturn, which is indicated by the gray background color.

The final note in this section is made on the potential pitfalls in the instrumental detection of lightning. Not every detected short optical or radio burst must be due to lightning. There are cosmic ray hits, adventitious light, or "hot pixels" of CCDs that could be misinterpreted as lightning. Switching operations in the radio receiver itself or other spacecraft instruments can cause radio bursts that mimic lightning, and electromagnetic cleanliness is always an issue. Even Jovian radio bursts observed from far distances can be misinterpreted as lightning [Fischer et al., 2006]. Therefore, care and caution is needed in the interpretation of bursty optical and radio data. The scientist has to be familiar with the instrument he/she is using and has to be aware that the instrument has a certain noise level, and this information is mostly not available for someone from outside. In this respect, the reader of the scientific article finally has to rely on the judgement of the investigating scientist, and so the validity of observations performed some decades ago is difficult to judge from today's point of view.

Table 1. Detection of signals attributed to lightning on various planets and Saturn's moon Titan. A yes with a question mark indicates scarce and/or ambigous observations. The gray background indicates combined optical and radio observations. VLF can be whistlers or sferics with a frequency below the ionospheric cutoff frequency

	optical	VLF	HF radio
Venus	Yes?	Yes?	Yes?
Earth	Yes	Yes	Yes
Jupiter	Yes	Yes	No
Saturn	Yes	Yes?	Yes
Titan	No	Yes?	No
Uranus	No	No	Yes
Neptune	No	Yes	Yes?

6. Past and Future Investigation of Terrestrial and Extraterrestrial Lightning

The detection of extraterrestrial lightning with space missions is mostly just a by-product of other investigations. Cameras on spacecraft have long lists of potential observing targets and lightning is just one of them, and not a priority as such in most cases. Radio instruments are usually optimized for the investigation of magnetospheric plasma waves or radio emissions. The upper frequency limit of 40 MHz of the Voyager PRA instrument was chosen because Jovian decametric radio emissions extend that far, and luckily enabled the detection of Saturn Electrostatic Discharges which of course were unknown at that time. The Japanese Lightning and Airglow Camera (LAC) onboard the Venus Climate Orbiter [Takahashi et al., 2008] is probably only the second instrument so far which is primarily dedicated to the investigation of extraterrestrial lightning. The first one was the Groza (Russian for "thunder") instrument onboard the Venera landers that provided probably the best indication of lightning on Venus so far.

6.1. Investigation of Terrestrial Lightning from Space

The investigation of terrestrial lightning is increasingly relying on detectors placed on Earth-orbiting satellites. Examples are the Optical Transient Detector (OTD) onboard the MicroLab-1 satellite that was able to image lightning even on the dayside [Christian et al., 1996], the Lightning Imaging Sensor (LIS) onboard the Tropical Rainfall Measuring Mission [Christian et al., 1999], or optical detectors placed onboard the Space Shuttle. The HF/VHF radio component of lightning has also been observed with some satellites like RAE-1 [Herman et al., 1965], ISS-b [Kotaki and Katoh, 1983], or Alexis, and with the latter the Blackbeard receiver detected the so-called Transionospheric Pulse Pairs [Massey and Holden,

1995]. Combined optical and radio observations were first realized by the FORTE satellite that was launched in August 1997 [Suszcynsky et al., 2000]. The radio frequency payload of the FORTE satellite consists of two broadband receivers capable to cover a 22 MHz band in the frequency range from 30-300 MHz [Jacobson et al., 1999], and a third receiver with an analog bandwidth of 85 MHz which digitizes with a sampling rate of 300 MHz resulting in a temporal resolution of a few nanoseconds [Light et al., 2001]. The receivers can be triggered autonomously or by the optical detector to record power time series with a duration of 8 ms. The typical duration of a terrestrial stroke is less than 0.1 ms, and in this way each stroke of a flash is recorded separately along with its optical pulse that usually lasts a few milliseconds. Furthermore, there is the Demeter spacecraft on a polar circular orbit at an altitude of 700 km. Its radio instrument goes from DC up to 3.5 MHz, and it has measured various types of radio emissions related to thunderstorm activity, most notably whistlers [Parrot et al., 2008]. The future Taranis mission [Lefeuvre et al., 2008] is probably the first satellite project dedicated solely to lightning and its effects, because it should mainly study the physics of TLEs and TGFs (Terrestrial Gamma Ray Flashes) from a polar orbit around Earth. Taranis will consist of a suite of instruments capable to detect the low and high frequency radio emissions, light, X-rays and gamma rays, and even energetic electrons related to lightning discharges. Such a multi-instrument and multi-phenomenology approach should open up new ways to detect and investigate lightning also at other planets.

6.2. Future Missions and the Dual-Phenomenology Approach for Extraterrestrial Lightning Investigations

The future Juno spacecraft will go into a polar orbit around Jupiter in 2016. In principle its radio instrument WAVES could measure also a lot of Jovian whistlers, but the main scientific objective is investigating the radio and plasma waves in Jupiter's polar magnetosphere. Similarly to FORTE, high temporal resolution observations of HF/VHF pulses of extraterrestrial lightning could reveal information about the substructure of the discharge process, where not much is known until now. Observations of SEDs with a temporal resolution of the order ~0.1 ms were already performed by the Voyagers and showed no clear sub-stroke structure but HF amplitude fluctuations of the order of tens of ms as well as less than 1 ms [Evans et al., 1983]. Cassini/RPWS is equipped with a similar millisecond mode. However, given the fact that terrestrial strokes last only of the order of 0.1 ms (100 μs) or even less, the temporal resolution used so far was probably not enough to resolve also the strokes of extraterrestrial flashes. A nanosecond time resolution like FORTE wouldn't be necessary and produce too much data volume, but an adequate temporal resolution for this task should be 1 μs with a record duration of about half a second. A triggering mechanism should be employed to avoid empty records. Such a system could as well be used to acquire waveforms of other magnetospheric radio emissions like kilometric radiation and study their fine structure. Future radio instruments onboard spacecraft to various planets (especially Saturn and Uranus where strong HF radio components are known to be present and maybe Neptune) should take this into account. Furthermore, radio wave direction-finding systems could as well help in localizing the radio source besides optical cameras looking out for convective cloud systems or direct flashes of light. Due to the short nature of the lightning radio signal such a receiver system needs to measure 3 autocorrelations and 6 crosscorrelations between 3 antennas in parallel, and a 12-bit ADC would be necessary to perform a radio wave direction-finding with sufficient accuracy.

As we have pointed out, at least a dual-phenomenology approach with optical and radio emissions should be applied at other planets for a reliable detection and investigation of lightning. This is in line with Desch et al. [2002] who stated that progress in resolving the controversy of Venus lightning would require either an extensive ground-based optical search, or an orbiter with a camera and HF receiver. Having the supremacy of optical detection in mind, the Japanese Lightning and Airglow Camera (LAC) will search for optical signals of lightning in the atmosphere of Venus starting in late 2010. The LAC rather uses an approach with high temporal resolution but low spatial resolution to detect lightning flashes [Takahashi et al., 2008]. The only drawback seems to be that no corresponding radio instrument is onboard the Venus Climate Orbiter (Akatsuki), and the controversy of Venus lightning mainly results from the single-instrument approaches. Therefore, combined optical observations by the LAC at Venus and by ground-based telescopes at Earth would be desirable. This would not be a dual- phenomenology but at least a dual-instrument approach. Even if optical detection of Venus lightning will be successful, it would be interesting to further investigate the nature of the signals measured by the PVO plasma wave instrument with its four channels. Clearly, an instrument with improved frequency resolution (more channels) would be required. The night side ionosphere seems to be a highly dynamic region with ion troughs and wave activity worth to be investigated in more detail. Wideband data of these whistler-mode emissions could show if their spectrum is similar to saucer emissions or lightning whistlers and in the latter case their dispersion could be determined providing the total electron content along the propagation path. A survey of HF emissions as stated by Desch et al. [2002] would as well be desirable, but one has to keep in mind that besides Solar radio bursts and potential HF emissions from Venus lightning no other wave activity is measurable close to Venus above ~1 MHz. Since data rates are always an issue in space missions, such a search would have to be limited to a few frequency channels and a trigger mechanism could further reduce the amount of data. An HF radio survey has the advantage of being able to detect lightning on the night and day side of Venus.

6.3. Laboratory Simulations

The recent advances in simulating extraterrestrial lightning in the laboratory are summarized in Yair et al. [2008]. Most experiments are based on sparks produced by high voltage electrodes or a laser induced plasma inside a vessel with a proper gas mixture. Most of these experiments are carried out at room temperature and at ambient or lower pressure. The simulation of relevant temperature ranges or higher pressures expected to be prevalent for lightning at the Jovian planets has hardly been realized due to high experimental efforts and costs. Modelling work is cheaper, but also in this area still much can be done. Cloud electrification models have been done for Jupiter [Yair et al., 1995], Neptune [Gibbard et al., 1999], Titan [Tokano et al., 2001], and recently for Venus [Michael et al., 2009]. In some of those models cloud convection is already included to model the complete thundercloud, but models specially devoted to convective cloud systems in extraterrestrial atmospheres, like e.g. Hueso and Sánchez-Lavega [2004] for Saturn, are also rare.

6.4. Ground-Based Radio Observations

Ground-based detection of HF radio emissions from extraterrestrial lightning with giant radio telescopes also has some potential for the future. The first detection of SEDs was made by Konovalenko et al. [2009] with the giant Ukraine UTR-2 radio telescope, and according to Zarka et al. [2004] the ground-based detection of lightning from Venus or Uranus [Zarka and Pedersen, 1986] might be possible as well with the growing Low Frequency Array (LOFAR). In the near future such ground-based observations could easily acquire waveform data.

6.5. Summary of Current and Future Space Mission for Lightning Investigation

To conclude, these are the most important current and future spacecraft missions that will investigate extraterrestrial lightning: The next highlight will be the Japanese Venus Climate Orbiter with its lightning camera that should start its investigation already in late 2010. Aplin [2006] has pointed out that an instrument to study atmospheric electricity at Mars is overdue. Cassini will continue to investigate the Saturnian system until 2017, and at the end of the mission we will have monitored the lightning activity for almost half a Saturnian year potentially allowing us to discern a seasonal influence on lightning activity. Its late orbit should bring the spacecraft between Saturn's atmosphere and its innermost ring, which might enable it to observe lightning storms from very close distances or detect more whistlers. No future missions are planned to Uranus and Neptune right now, but the Jovian system should be investigated by Juno in 2016 and in the 2020s by the joint EJSM (Europa Jupiter System Mission) that should be equipped with radio as well as optical instruments. Takahashi et al. [2009] proposed to carry a special lightning camera on the planned European Jupiter Ganymede Orbiter similar to the one used at Venus.

Acknowledgment

G. Fischer was supported by a grant (P21295-N16) from the Austrian Science Fund (FWF).

References

Akalin, F. et al. (2006). First whistler observed in the magnetosphere of Saturn, *Geophys. Res. Lett.*, *33*, L20107.

Aplin, K. L. (2006). Atmospheric electrification in the solar system, *Surveys in Geophys.*, *27*, 63-108.

Baines, K. H. et al. (2007). Polar lightning and decadal-scale cloud variability on Jupiter, *Science*, *318*, 226-229.

Belton, M. J. S. et al. (1991). Images from Galileo of the Venus cloud deck, *Science*, *253*, 1531-1536.

Borucki, W. J. & Magalhães, J. A. (1992). Analysis of Voyager 2 images of jovian lightning, *Icarus*, *96*, 1-14.

Borucki, W. J. (1982). Comparison of Venusian lightning observations, *Icarus*, *52*, 354-364.

Borucki, W. J. et al. (1982). Lightning activity on Jupiter, *Icarus*, *52*, 492-502.

Christian, H. J. et al. (1996). *The Optical Transient Detector (OTD)*, Proc. 10th Int. Conf. on Atmospheric Elec., Osaka, Japan, 368-371.

Christian, H. J. et al. (1999). *The Lightning Imaging Sensor*, Proc. 11th Int. Conf. on Atmospheric Elec., Guntersville, Alabama, 746-749.

Cook, A. F. et al. (1979). First results on jovian lightning, *Nature*, *280*, 794-795.

Desch, S. J. et al. (2002). Progress in planetary lightning, *Rep. Prog. Phys.*, *65*, p.955.

Dubrovin, D. et al. (2010). Sprite discharges on Venus and Jupiter-like planets: a laboratory investigation, *J. Geophys. Res.*, *115*, A00E34.

Dyudina, U. A. et al. (2004). Lightning on Jupiter observed in the Hα line by the Cassini imaging science subsystem, *Icarus*, *172*, 24-36.

Dyudina, U. A. et al. (2007). Lightning storms on Saturn observed by Cassini ISS and RPWS in the years 2004-2006, Icarus, *190*, 545-555.

Dyudina, U. A. et al. (2010). Detection of visible lightning on Saturn, *Geophys. Res. Lett.*, *37*, L09205.

Esposito, L. W. et al. (1983). The clouds and hazes of Venus, in "*Venus*" edited by Hunten et al., 484-564.

Evans, D. R. et al. (1982). The source of Saturn electrostatic discharges, *Nature*, 299, 236-237.

Evans, D. R. et al. (1983). Saturn's Electrostatic Discharges: Properties and theoretical considerations, *Icarus*, *54*, 267-279.

Farrell, W. M. et al. (1999a). Detecting electrical activity from Martian dust storms, *J. Geophys. Res.*, *104*, 3795-3801.

Farrell, W. M. et al. (1999b). A model of the lightning discharge at Jupiter, *Geophys. Res. Lett.*, *26*, 2601-2604.

Fischer, G. et al. (2006). Discrimination between Jovian radio emissions and Saturn electrostatic discharges, *Geophys. Res. Lett.*, *33*, L21201.

Fischer, G. et al. (2007a). Nondetection of Titan lightning radio emissions with Cassini/ RPWS after 35 close Titan flybys, *Geophys. Res. Lett.*, *34*, L22104.

Fischer, G. et al. (2007b). Analysis of a giant lightning storm on Saturn, *Icarus*, *190*, 528-544.

Fischer, G. et al. (2008). Atmospheric electricity at Saturn, *Space Sci. Rev.*, *137*, 271-285.

Fulchignoni, M. et al. (2005). In situ measurements of the physical characteristics of Titan's environment, *Nature*, *438*, 785-791.

Gibbard, S. G. et al. (1999). Lightning on Neptune, *Icarus*, *139*, 227-234.

Gurnett, D. A. & Frank, L. A. (1972). VLF hiss and related plasma observations in the polar magnetosphere, *J. Geophys. Res.*, *77*, 172-190.

Gurnett, D. A. et al. (1979). Whistlers observed by Voyager 1 – Detection of lightning on Jupiter, *Geophys. Res. Lett.*, *6*, 511-514.

Gurnett, D. A. et al. (1990). Whistlers in Neptune's magnetosphere: Evidence of atmospheric lightning, *J. Geophys. Res.*, *95*, 20967-20976.

Gurnett, D. A. et al. (1991). Lightning and plasma wave observations from the Galileo flyby of Venus, *Science*, *253*, 1522-1525.

Gurnett, D. A. et al. (2001). Non-detection at Venus of high-frequency radio signals characteristic of terrestrial lightning, *Nature*, *409*, 313-315.

Gurnett, D. A. et al. (2004). The Cassini Radio and Plasma Wave Science investigation, *Space Sci. Rev.*, *114*, 395-463.

Gurnett, D. A. et al. (2010). Non-detection of impulsive radio signals from lightning in Martian dust storms using the radar receiver on the Mars Express spacecraft, *Geophys. Res. Lett.*, *37*, L17802.

Hamelin, M. et al. (2009). Comment on "Evidence of electrical activity on Titan drawn from the Schumann resonances sent by Hugens probe", *Icarus*, *204*, 349-351.

Hansell, S. A. et al. (1995). Optical detection of lightning on Venus, *Icarus*, *117*, 345-351.

Herman, J. R. et al. (1965). Radio astronomy explorer (RAE-1) observations of terrestrial radio noise, *Planet. Space Sci.*, *21*, 443-461.

Holzworth, R. H. et al. (1999). Lightning whistler waves in the high-latitude magnetosphere, *J. Geophys. Res.*, *104*, 17369-17378.

Huba, J. D. (1992). Theory of small-scale density and electric field fluctuations in the nightside Venus ionosphere, *J. Geophys. Res.*, *97*, 43-50.

Hueso, R. & Sánchez-Lavega, A. (2004). A three-dimensional model of moist convection for the giant planets, II: Saturn's water and ammonia moist convective storms, *Icarus*, *172*, 255-271.

Hueso, R. et al. (2008). The atmosphere of Venus: winds and clouds observed by VIRTIS/Venus Express, *Lecture Notes and Essays in Astrophys.*, *3*, 1-12.

Hunten, D. M. (1995). Venus lightning: pros and cons, *Adv. Space Res.*, *15*, (4)109-(4)112.

Jacobson, A. R. et al. (1999). FORTE observations of lightning radio-frequency signatures: Capabilities and basic results, *Radio Sci.*, *34*, 337-354.

Kaiser, M. L. et al. (1983). Atmospheric storm explanation of saturnian electrostatic discharges, *Nature*, *303*, 50-53.

Kaiser, M. L. et al. (1984). Saturn's ionosphere-inferred electron densities, *J. Geophys. Res.*, *89*, 2371-2376.

Kaiser, M. L. et al. (1991). Restrictions on the characteristic of Neptunian lightning, *J. Geophys. Res.*, *96*, 19043-19047.

Konovalenko, A. et al. (2009). Ground-based decametre wavelength observations of Saturn's lightning during the giant storm detected by Cassini spacecraft, Icarus, submitted.

Kotaki, M. & Katoh, C. (1983). The global distribution of thunderstorm activity observed by the ionosphere sounding satellite (ISS-b), *J. Atmos. Terr. Phys.*, *45*, 833.

Krasnopolsky, V. A. (1983). Venus spectroscopy in the 3000-8000 Å region by Veneras 9 and 10, in "*Venus*" edited by Hunten et al., 459-483.

Ksanfomaliti, L. V. (1980). Discovery of frequent lightning discharges in the clouds on Venus, *Nature*, *284*, 244-246.

Ksanfomaliti, L. V. et al. (1979). Electrical discharges in the atmosphere of Venus, *Sov. Astron. Lett.*, *5*, 122-126.

Ksanfomaliti, L. V. et al. (1982). The low-frequency electromagnetic field in the Venus atmosphere: evidence from Venera 13 and Venera 14, Sov. *Astron. Lett.*, *8*, 230-232.

Ksanfomaliti, L. V. et al. (1983). The electrical activity of the atmosphere of Venus, in "Venus" edited by Hunten et al., 565-603.

Lanzerotti, L. J. (1996). Radio frequency signals in Jupiter's atmosphere, *Science*, *272*, 858-860.

Lefeuvre, F. et al. (2008). TARANIS-A satellite project dedicated to the physics of TLEs and TGFs, *Space Sci. Rev.*, *137*, 301-315.

Light, T. E. et al. (2001). Coincident radio frequency and optical emissions from lightning, observed with the FORTE satellite, *J. Geophys. Res.*, *106*, 28223-28231.

Little, B. et al. (1999). Galileo images of lightning on Jupiter, *Icarus*, *142*, 306-323.

Maeda, K. & Grebowsky, J. M. (1989). VLF emission bursts in the terrestrial and venusian nightside troughs, *Nature*, *341*, 219-221.

Majeed, T. & McConnell, J. C. (1996). Voyager electron density measurements on Saturn: Analysis with a time dependent ionospheric model, *J. Geophys. Res.*, *101*, 7589-7598.

Markiewicz, W. J. (2007). Morphology and dynamics of the upper cloud layer of Venus, *Nature*, *450*, 633-636.

Massey, R. S. & Holden, D. N. (1995). Phenomenology of transionospheric pulse pairs, *Radio Sci.*, *30*, 1645-1659.

Melnik, O. & Parrot, M. (1998). Electrostatic discharges in Martian dust storms, *J. Geophys. Res.*, *103*, 29107-29117.

Michael, M. et al. (2009). Highly charged cloud particles in the atmosphere of Venus, *J. Geophys. Res.*, *114*, E04008.

Morente, J. A. et al. (2008). Evidence of electrical activity on Titan drawn from the Schumann resonances sent by Huygens probe, *Icarus 195*, 802-811.

Parrot M. et al. (2008). DEMETER observations of EM emissions related to thunderstorms, *Space Sci. Rev. 137*, 511-519.

Peratt A. L. & Dessler, A. J. (1988). Filamentation of volcanic plumes on the Jovian satellite Io, *Astrophys. and Space Sci. 144*, 451-461.

Porco C. C. et al. (2004). Cassini imaging science: Instrument characteristics and anticipated scientific investigations at Saturn, *Space Sci. Rev. 115*, 363-497.

Porco C.C. et al. (2005). Cassini imaging science: Initial results on Saturn's atmosphere, *Science 307*, 1243-1247.

Rakov V. A. & Uman, M. A. (2003). *Lightning, Physics and Effects*, Cambridge Univ. Press, Cambridge.

Rinnert K. et al. (1998). Measurements of radio frequency signals from lightning in Jupiter's atmosphere, *J. Geophys. Res. 103*, 22979-22992.

Ruf C. et al. (2009). Emission of non-thermal microwave radiation by a Martian dust storm, *Geophys. Res. Lett. 36*, L13202.

Russell C.T. (1991). Venus lightning, *Space Sci. Rev. 55*, 317-356.

Russell C.T. (1993). Planetary lightning, *Annu. Rev. Earth Planet. Sci. 21,* 43-87.

Russell, C. T. et al. (1989). VLF bursts in the night ionosphere of Venus – Estimates of the Poynting flux, *Geophys. Res. Lett.*, *16*, 579-582.

Russell, C. T. et al. (2007). Lightning on Venus inferred from whistler-mode waves in the ionosphere, *Nature*, *450*, 661-662.

Russell, C. T. et al. (2008). Whistler mode waves from lightning on Venus: Magnetic control of ionospheric access, *J. Geophys. Res.*, *113*, E00B05.

Sagdeev, R. Z. et al. (1986). Overview of VEGA Venus balloon in situ meteorological measurements, *Science*, *231*, 1411-1414.

Scarf, F. L. & Russell, C. T. (1988). Evidence of lightning and volcanic activity on Venus– Pro and con, *Science*, *240*, 222-224.

Scarf, F. L. (1986). Comment on "Venus nightside ionospheric troughs – Implications for evidence of lightning and volcanism", *J. Geophys. Res.*, *91*, 4594-4598.

Scarf, F. L. et al. (1980). Lightning on Venus: Orbiter detection of whistler signals, *J. Geophys. Res.*, *85*, 8158-8166.

Shao, X. M. & Krehbiel, P. R. (1996). The spatial and temporal development of intracloud lightning, *J. Geophys. Res.*, *101*, 26641-26668.

Singh, R. N. & Russell, C. T. (1986). Further evidence for lightning on Venus, *Geophys. Res. Lett.*, *13*, 1051-1054.

Smith, B. A. et al. (1979). The Jupiter system through the eyes of Voyager 1, *Science*, *204*, 951-972.

Strangeway, R. J. (1992). An assessment of lightning or in situ instabilities as a source for whistler mode waves in the night ionosphere of Venus, *J. Geophys. Res.*, *97*, 12203-12215.

Suszcynsky, D. M. et al. (2000). FORTE observations of simultaneous VHF and optical emissions from lightning: Basic phenomenology, *J. Geophys. Res.*, *105*, 2191-2201.

Takahashi, Y. et al. (2009). Optical lightning detector for Jovian orbiter: OLD, talk presented at the European Planetary Science Congress in Potsdam, Germany.

Takahashi, Y. et al., (2008). Lightning detection by LAC onboard the Japanese Venus Climate Orbiter, Planet-C, *Space Sci. Rev.*, *137*, 317-334.

Taylor, H. A. & Cloutier, P. A. (1986). Venus-Dead or alive? *Science*, *234*, 1087-1093.

Taylor, H. A. & Cloutier, P. A. (1988). Evidence of lightning and volcanic activity on Venus: Pro and con: Response, *Science*, *240*, 224-226.

Taylor, H. A. & Cloutier, P. A. (1992). Non-evidence of lightning and associated volcanism at Venus, *Space Sci. Rev.*, *61*, 387-391.

Taylor, H. A. et al. (1985). Venus nightside ionospheric troughs: Implications for evidence of lightning and volcanism, *J. Geophys. Res.*, *90*, 7415-7426.

Taylor, H. A. et al. (1986). Reply, *J. Geophys. Res.*, *91*, 4599-4605.

Taylor, H. A. et al. (1987). Venus "lightning" signals reinterpreted as in situ plasma noise, *J. Geophys. Res.*, *92*, 9907-9919.

Taylor, W. W. L. et al. (1979). Evidence for lightning on Venus, *Nature*, *279*, 614-616.

Thomas, R. J. et al. (2007). Electrical activity during the 2006 Mt. Saint Augustine volcanic eruptions, *Science*, *315*, 1097.

Tokano, T. et al. (2001). Modelling of thunderclouds and lightning generation on Titan, *Planet. Space Sci.*, *55*, 1978-1989.

Van der Velde, O. A. et al. (2006). Observations of the relationship between sprite morphology and in-cloud lightning processes, *J. Geophys. Res.*, *111*, D15203.

Warwick, J. W. et al. (1981). Planetary radio astronomy observations from Voyager 1 near Saturn, *Science*, *215*, 582-587.

Williams, M. A. et al. (1982). The transmission to space of the light produced by lightning in the clouds of Venus, *Icarus*, *52*, 166-170.

Yair, Y. et al. (1995). Lightning generation in a Jovian thundercloud: Results from an axisymmetric numerical cloud model, *Icarus*, *115*, 421-434.

Yair, Y. et al. (2008). Updated review of planetary atmospheric electricity, *Space Sci. Rev.*, *137*, 29-49.

Yair, Y. et al. (2009). A study of the possibility of sprites in the atmospheres of other planets, *J. Geophys. Res.*, *114*, E09002.

Zarka, P. & Pedersen, B. M. (1986). Radio detection of Uranian lightning by Voyager 2, *Nature*, *323*, 605-608.

Zarka, P. (1985). On the detection of radio bursts associated with Jovian and Saturnian lightning, *Astron. Astrophys.*, *146*, L15-L18.

Zarka, P. et al. (2004). Study of solar system planetary lightning with LOFAR, *Planet. Space Sci.*, *52*, 1435-1447.

In: Lightning: Properties, Formation and Types
Editor: Matthew D. Wood
ISBN: 978-1-61122-198-5

Chapter 3

RECENT STUDIES OF SCHUMANN RESONANCE AND ELF TRANSIENTS

***M. Hayakawa*[1], *A. P. Nickolaenko*[2], *A. V. Shvets*[2] *and Y. Hobara*[1]**

[1]Department of Electronic Engineering and Research Station on Seismo Electromagnetics, The University of Electro–Communications, Chofu, Tokyo 182-8585, Japan

[2]Usikov's Institute for Radio-Physics and Electronics, National Academy of Sciences of the Ukraine, 12 Acad. Proskura street, Kharkov 61085, Ukraine

ABSTRACT

We overview the results of Schumann resonance and ELF studies obtained during the last decade. The following items are addressed: ionospheric non-uniformities related to seismic activity, formal proofs of the resonance line splitting under the influence of geomagnetic field, impacts of solar proton events and extra-solar gamma-ray bursts on resonance parameters, ELF intensity and global land temperature, Schumann resonance on other planets, data from a new observatory at Spitzbergen, sprites and ELF transients, biological effects of Schemann resonance, solutions of the inverse problem, the 'terminator effect' and its controversial explanations, the accuracy of locating the Q-bursts sources, a comparison of waveforms of Q-bursts (experiment versus theory), parameters of thunderstorm activity which follow from the optical transient detector data.

Keywords: Schumann resonance, ELF transients, Ionospheric non-uniformities, Global thunderstorm activity, Inverse problem, Sprites

1. INTRODUCTION

A lightning discharge is known to radiate electromagnetic waves in a wide frequency range with a peak at ~10 kHz in the VLF (very low frequency) band. The radiation is called a sferic and depending on the frequency of interest we name VLF sferics and ELF (extremely low frequency) sferics. There are two different kinds of ELF sferics; (1) stationary Schumann resonances (SRs) and (2) non-stationary ELF transients. As compared with VLF sferics, the most important property of ELF sferics is their extremely low propagation loss in the Earth-ionosphere waveguide so that they can travel round the globe. SRs are a global resonance phenomena excited primarily by background lightning discharges in the Earth-ionosphere cavity and their resonance frequencies are 8, 14, 20, 26 Hz etc. (*Nickolaenko and Hayakawa*, 2002). These SRs were studied extensively in 1960s and 1970s (e. g., *Balser and Wagner*, 1960, 1962; *Polk and Fitchen*, 1962; *Rycroft*, 1963; *Jones*, 1964; *Madden and Thompson*, 1965; *Ogawa et al.*, 1966; *Sao et al.*, 1973), but the SR studies died completely in around 1980 because of losing its scientific significance. However, there has been an increasing interest again in this SR study during the last decade. We can list the following two important motivations for this interest. The first is that SR is suggested to be able to monitor the global warming as a global thermometer (*Williams*, 1992). The second reason is the finding of transient luminous events (TLEs); that is, optical emissions (red sprites etc.) in the mesosphere excited by a huge (mainly positive) lightning discharges were observed in 1990s (*Franz et al.*, 1990). Also, those TLEs are closely related with a non-stationary ELF transient (*Boccippio et al.*, 1995). Such intense ELF transients are currently utilized to study extensively those TLEs (e. g., *Fullekrug et al. (Eds)*, 2006).

We have published a reviewer on this topic in 2007 (*Nickolaenko and Hayakawa*, 2007), and the present review is an extensively updated version of our previous one. After the publication of *Nickolaenko and Hayakawa* (2007), there have been published two interesting documents ; one is a review on SRs by *Price et al.* (2007) and another is a special issue on SRs (*Pasko (Ed.)*, 2007) in which about 20 papers are included. When writing a review, it is impossible to have the complete coverage of all topics in ELF studies because of a limited space, so that this review is decisively based on our own choice and preference.

The majority of results published were based on the power spectra of different field components measured at different sites giving information on the lower ionosphere and on the global thunderstorm activity. We below deal with anomalous resonance signals observed prior to earthquakes (EQs), sudden changes of resonance parameters observed during the Solar Proton Events (SPE) and during the extra-solar gamma-ray bursts, and the regular seasonal changes in thunderstorms and their association with the ground temperature. It was demonstrated formally that the cause of elliptical polarization of the horizontal magnetic field vector is the line splitting of SR oscillations. There is a renewal of interest toward the extra-terrestrial SR studies, particularly, in connection with the Cassini/Huygens mission to Titan. We describe recent results on the inverse problem, mention some results of a new SR observatory at Spitzbergen, sprites and ELF transients, and biological effects of SRs, address the 'terminator effect' and its controversial explanations, discuss the accuracy of location of the Q-bursts sources and a comparison of experimental and theoretical waveforms of Q-busts, and finally demonstrate parameters of thunderstorm activity that follow from the optical transient detector (OTD) data.

2. EFFECTS CONNECTED WITH THE NON-UNIFORMITIES IN THE LOWER IONOSPHERE

2.1. Seismic Effects in the SR Band

The anomalous effects in SR were observed, for the first time, at Nakatsugawa observatory, Japan (Geographic coordinates: 35.4° N, 137.5° E) in association with an EQ in Taiwan (see e.g., *Hayakawa et al.,* 2005a). Figure 1 illustrates the dynamic spectrum of ELF waves there in September 1999. The Universal Time (UT) is shown on the abscissa in days (September 1999), the frequency is shown on the ordinate in the range from 0.1 to 50 Hz, and the intensity of electromagnetic field is shown by color. The anomaly is characterized by the enhancement of SR around the fourth harmonic (about 25 Hz) and the presence of pulsed signals at the same frequency. This anomalous Schumann effect appears a few days before the Chi-chi EQ on September 21, 1999.

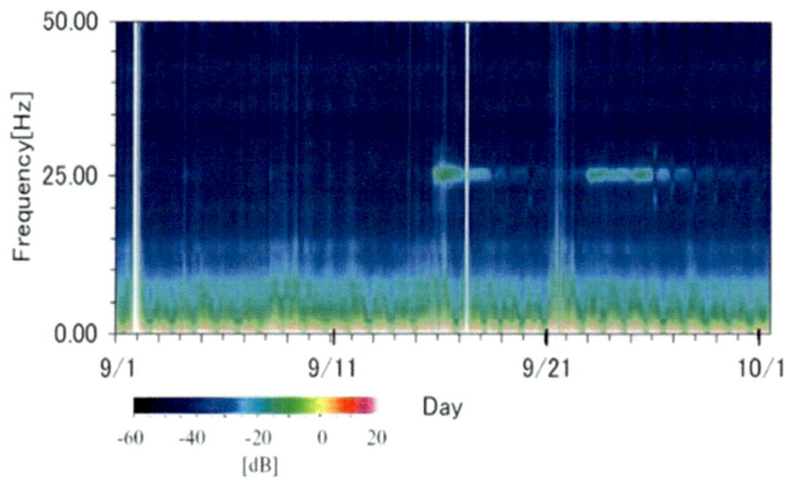

Figure 1. The anomalous increase in the fourth harmonic (around the frequency of 25 Hz) and pulsed unusual SR signals are observed in Japan associated with an EQ in Taiwan.

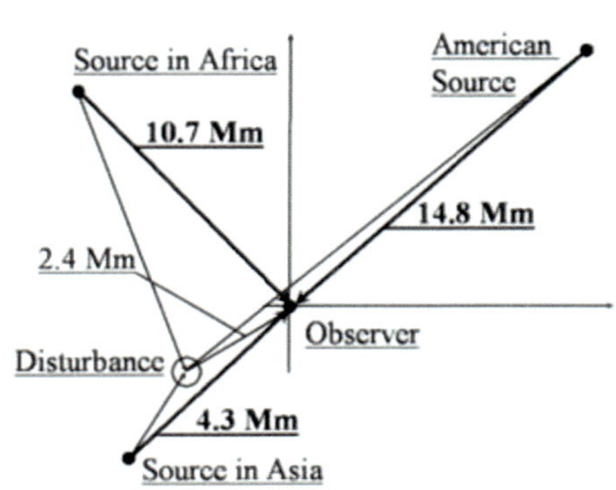

Figure 2. ELF propagation paths in the 'flat' projection. Shown are the characteristic positions and distances from global thunderstorm centers as seen from Nakatsugawa observatory.

A recent work by *Ohta et al.* (2006) has confirmed the same phenomena for many EQs over several years. They have found that all of seven inland EQs are accompanied with this kind of anomaly at Nakatsugawa. One possible explanation was suggested by *Hayakawa et al.* (2005a) and *Nickolaenko et al.* (2006a): the origin of such an anomaly is due to the ELF radio wave scattering by a severe modification of the lower ionosphere localized over the EQ. Figure 2 depicts the 'flat' geometry of the ELF radio propagation in the Earth – ionosphere cavity with a localized disturbance over the EQ. Shown are the characteristic positions and distances from three major thunderstorm centers: Africa, America, and Asia.

The radio propagation problem was solved in a non-uniform cavity by using the Stratton-Chu integral equation with the localized ionosphere modification over Taiwan (*Nickolaenko*, 1994; *Nickolaenko and Hayakawa*, 2002; *Nickolaenko et al.*, 2006a). The point vertical electric dipole source was placed at one of the following points: Indonesia (10°N, 105°E), Guinea Gulf (0°N, 25°E), or in Amazon basin (10°N, 60°W). The relevant results are depicted in Figure 3. The right plot in this figure presents the frequency dependence of the normalized disturbance of E_Z field. The plots for different sources have rather complicated patterns, but the American source combined with the ionosphere irregularity over Taiwan causes an increase in the field amplitude (up to 50%) at particular frequencies.

The left panels in Figure 3 show the power spectra, and the computations confirm that the effect must be expected around the fourth SR frequency when the thunderstorms occur in the South America. A conclusion was made that our experimental result might be explained by the wave interference between the direct signal from South America and the ELF wave scattered back from Taiwan in the above geometry.

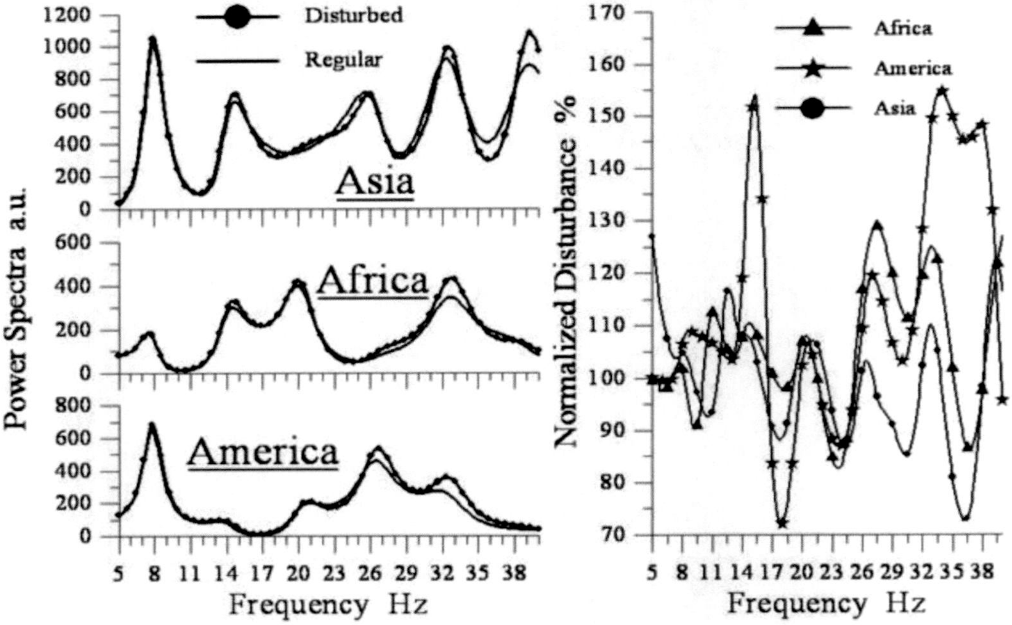

Figure 3. Model SR spectra and normalized field modification computed for the point source in Asia (10°N, 105°E), Africa (0°N, 25°E), or America (10°N, 60°W).

The model of a single point source is unrealistic, and so computations were extended for the global distribution of lightning strokes. The latter was adopted from the data collected by the spacecraft (*Christian et al.*, 2003). The UT was fixed, and the relevant OTD (optical transients detector) map was adopted of the annually averaged global distribution of lightning strokes (see *Hayakawa et al.*, 2005a; *Nickolaenko et al.*, 2006a for details). It was assumed that individual lightning strokes form the Poisson succession, so that the power spectra are summed over the cells of the global OTD map. Figure 4 presents the power spectra in the regular and disturbed Earth–ionosphere cavities for four characteristic times: UT=3, 9, 15, and 21 hr. The three latter times correspond to the moments when the global lightning activity reaches its maximum in Asia (9 hr UT), Africa (15 hr UT), or in America (21 hr UT). The UT=3 hr is relevant to the daily lowest level of global thunderstorms. Figure 4 indicates that global thunderstorms combined with the localized non-uniformity of the lower ionosphere may increase the power at the fourth SR by about 20%. Thus, the mechanism suggested qualitatively explains the experimental results, though there is still some gap in intensity between the observation and theory.

An additional experimental fact has also been observed at a different Japanese station of Moshiri in Hokkaido (47N°, 142°E) in a possible association with another Ping-tong EQ in Taiwan (*Hayakawa et al.*, 2008). The general behaviors of anomalous SRs at Moshiri are found to be consistent with our former works, and we have found again that the observed SR anomaly can be well explained in terms of the interference between the direct ELF wave and that scattered (or reflected) by the seismo-ionospheric disturbance over Taiwan.

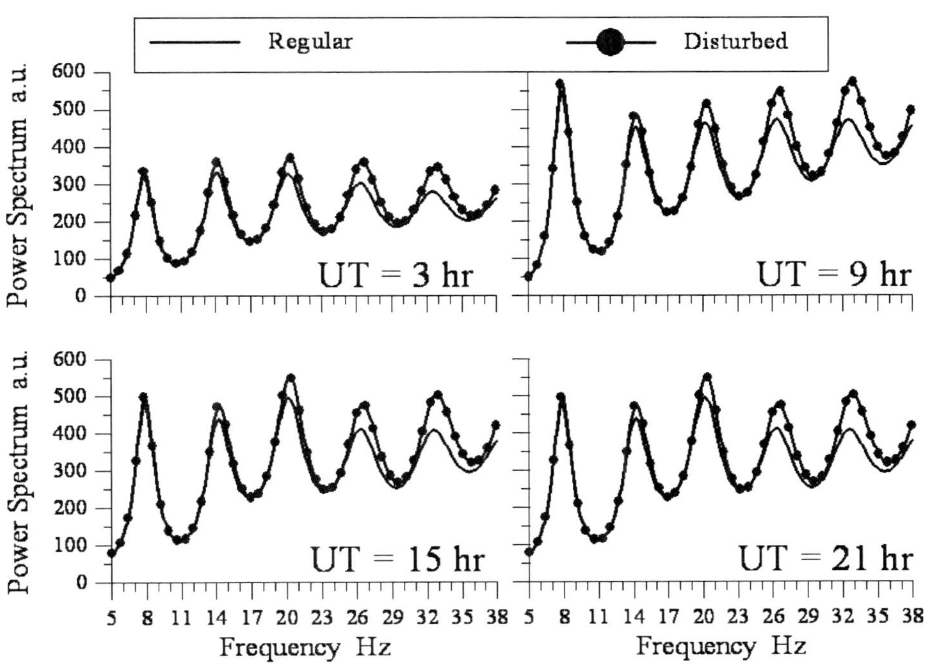

Figure 4. The regular and disturbed spectra at Nakatsugawa observatory for particular UT times.

2.2. Formal Proofs of the Detection of the SR Line Splitting

The formal treatment and model computations were made and compared with the results of modern SR monitoring demonstrating that the line splitting really exists (*Nickolaenko et al.*, 2002, 2004). The diagrams in Figure 5 explain the lift of the oscillations degeneracy. The splitting is caused by the anisotropy of the ionospheric plasma embedded in the geomagnetic field. As a result, three different eigenfunctions of the first SR acquire different azimuth numbers: $m = -1$, 0, and +1. The function with $m = -1$ is the radio wave traveling from west to east along the equator (see the upper diagram in Figure 5). The function with $m = 0$ is the standing wave. The wave travels from east to west when $m = +1$. All of them have different complex eigenvalues and frequency characteristics shown in the lower part of Figure 5. The line splitting cannot be seen simply in the power spectra of a field component, e.g., the power spectrum of the H_φ magnetic field remains a smooth and uni-modal function of frequency as is shown in Figure 5. In Figure 5 we give the expressions to the resonance denominators $|R_{n,m}|^{-2}$ (see *Nickolaenko and Hayakawa*, 2002). n=1 means the first resonance mode, and the azimuth index m=−1, 0, and +1. We here cite the expressions for the resonance denominators by using the component of the tensor surface impedance $Z = \begin{bmatrix} Z_1 & Z_2 \\ -Z_2 & Z_1 \end{bmatrix}$, the effective ionospheric height (h) and angular wave frequency (ω). The value of ω_1, denotes the first ideal SR frequency, and c, the velocity of light.

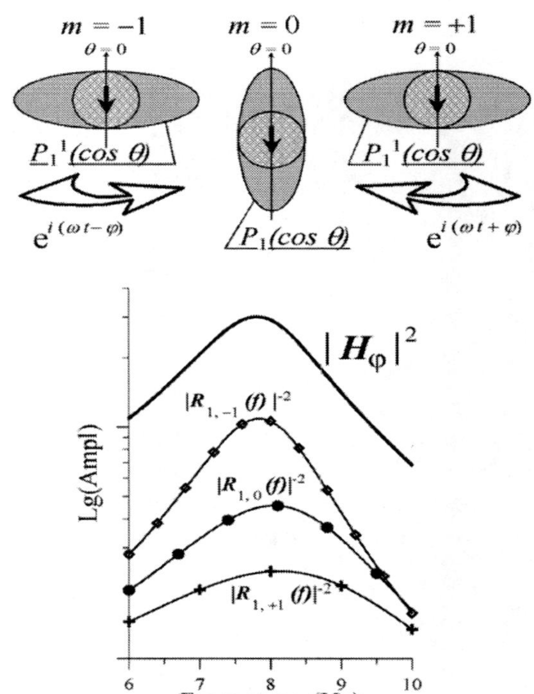

Figure 5. Eigen-values split in the presence of the geomagnetic field (upper panel). The lower diagram shows the eigenfunctions at the first SR mode and relevant frequency responses.

$$
R_{n,m} = \begin{cases} \omega_1^2 - \omega^2 + \dfrac{ic\omega Z_1}{h} & m=0 \\ \omega_1^2 - \omega^2 + \dfrac{ic\omega Z_1}{h} + \dfrac{3c\omega Z_2}{4h} & m=+1 \\ \omega_1^2 - \omega^2 + \dfrac{ic\omega Z_1}{h} - \dfrac{3c\omega Z_2}{4h} & m=-1 \end{cases}
$$

In order to detect the line splitting, *Sentman* (1987) suggested exploiting the polarization of the horizontal magnetic field vector. He performed measurements and concluded that the frequency dependence of the polarization sense indicates the line splitting, although formal treatment of the problem and model computations were not performed at that time. This job was done lately (*Nickolaenko et al.*, 2002, 2004), and the model data were then compared with the experiment on the modern equipment. The reciprocity is demonstrated of the model and experimental data in Figure 6. In the figure the polarization is expressed in color (- indicates the left-handed polarization, and +, right-handed polarization). Computations show that the line splitting is observed as the positive polarization sense at the resonance frequencies, and the line splitting is seen as a constant dark blue line around 8 Hz and at 14Hz.

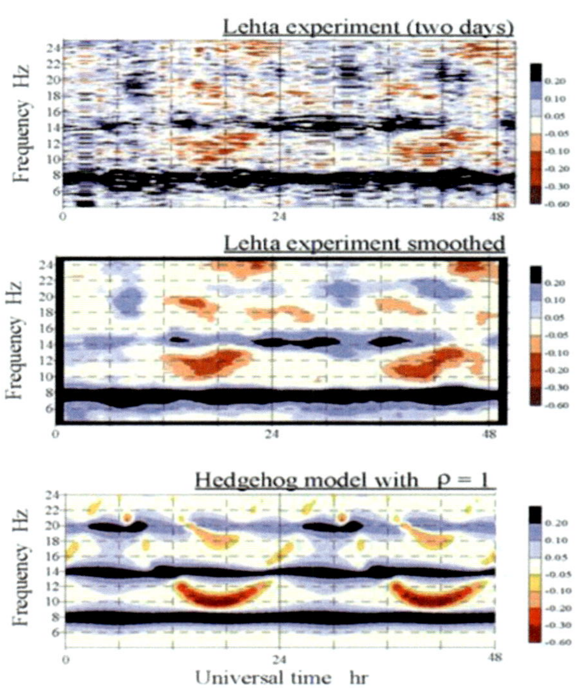

Figure 6. Line splitting in the spectra of wave polarization: theory (bottom panel) and experiment (upper two panels). The sense of polarization is given in color (+: right-handed and -, left-handed).

$$\omega_1^2 - \omega^2 + \frac{ic\omega Z_1}{h} m=0.$$

The similar behavior of wave polarization was observed at remote sites, Lehta (64°N, 34°E) and Karymshino (53°N, 158°E). A comparison with the model data showed that the anisotropy parameter is $\omega_H/\nu \approx 0.5$, where ω_H is the electron gyrofrequency of the ionosphere and ν is its effective electron collision frequency. It is interesting to mention that the patterns similar to those of model computations are also observed in the wave polarization at the Alaska observatory (*D. Sentman*, private communication).

2.3. The Impact of Solar Proton Events

Solar proton events (SPEs) are catastrophic bursts in the solar particle flow when relativistic protons reach the Earth. The interaction of the geomagnetic field with charged particles is complicated, and it is usually followed by the proton precipitation from the Earth's radiation belts. As a result, symmetric modifications occur in the polar ionospheric areas regarded as polar cap absorption (PCA) being the radio blackout in the band from ELF to the short waves.

The ionospheric modification may cause an abrupt change of the SR frequency sometimes combined with an increase in the signal intensity. Two SPEs are shown in Figs.7 and 8. The first one in Figure 7 is the Bastilia Day SPE on July 14, 2000, which was detected simultaneously at two sites: Lehta and Karymshino. Two upper graphs in Figure 7 depict variations of peak frequencies of the first and the second SR modes (F1 and F2) at Lehta (E_Z component). The lower plots show alterations of the first mode frequency in the orthogonal horizontal magnetic field components at Karymshino: D–component (approximately east–west) and H–component (approximately south–north). The vertical bar denotes the onset of the SPE.

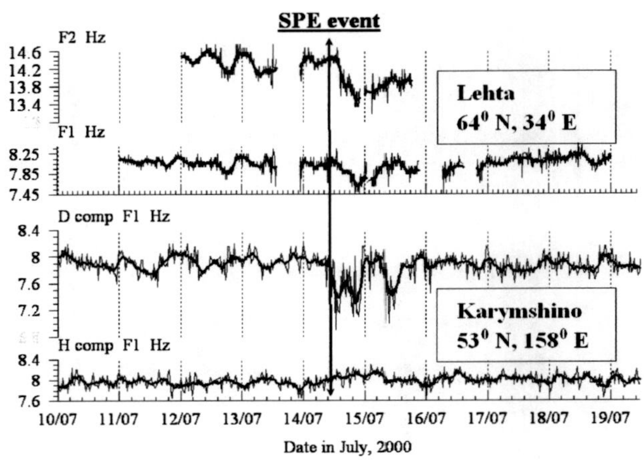

Figure 7. SPE on July 14, 2000: a worldwise decrease in peak frequency. F1 and F2 are the first and second SR modes. Hy (D component) and Hx (H component).

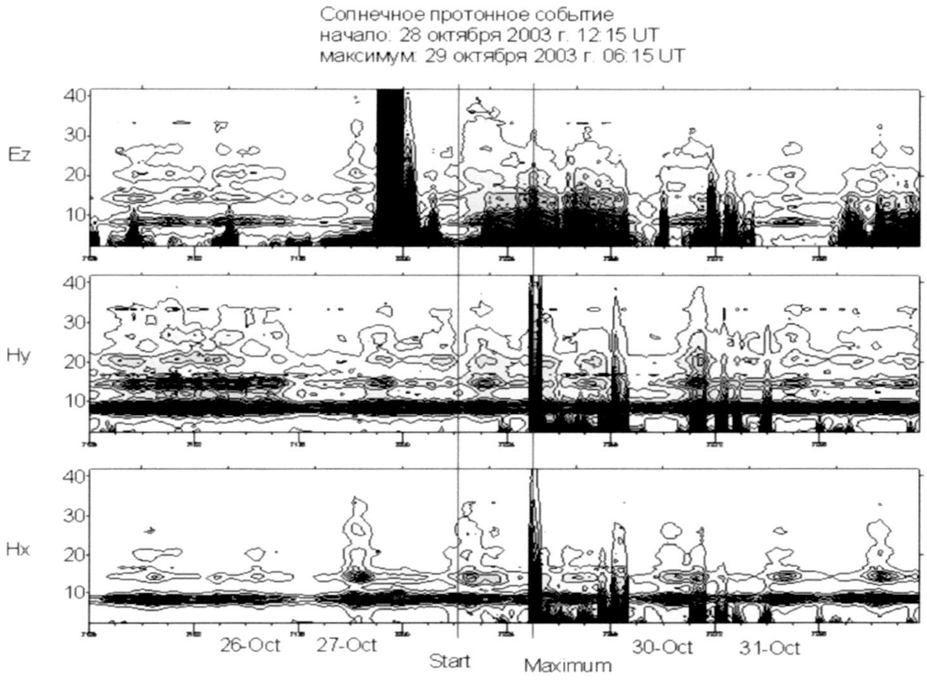

Figure 8. An increase in the ULF 'tail' and the relevant growth in the SR amplitude during Hallowing SPE of Oct. 29, 2003 (P_{max} = 29500). Hy (D component) and Hx (H component).

Figure 8 presents the sonagrams related to the Hallowing (October 29), 2003 SPE, which is adapted from *Shvets et al.* (2005). The dynamic spectra are shown here of all three field components recorded at Lehta. The first vertical line denotes the SPE onset. The familiar horizontal tracks correspond to the peaks in SR intensity. One may see that an increase in the geomagnetic activity followed with approximately one–day delay, and even the lower part of E_Z SR band was distorted by electromagnetic activity in the magnetosphere.

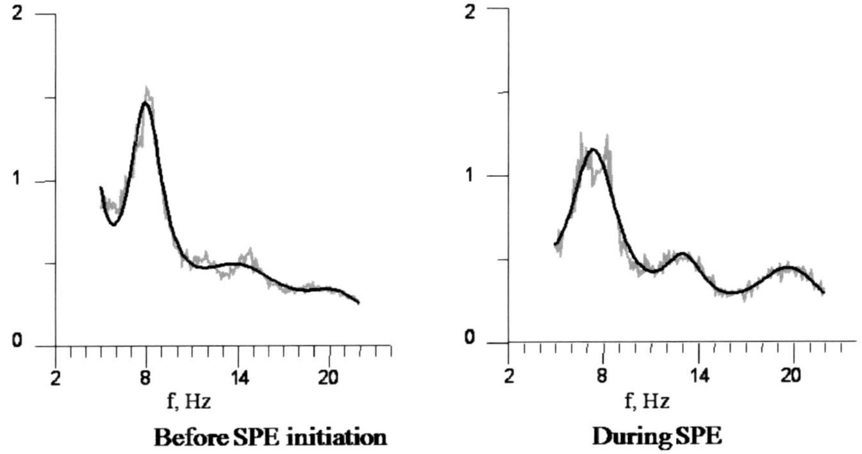

Figure 9. The fine structure of $|E_Z|$ spectrum during a SPE.

Figure 9 depicts the 'line splitting' observed during the SPE. The left panel depicts the SR spectrum recorded just before the SPE, and the right panel shows the corresponding spectrum of E_Z component after the SPE initiation. The original spectra with characteristic fine structures were fit by Lorentzian curves relevant to three sub-peaks and a 'tail' of the ULF signal. The peak separation in the right plot is too large to be attributed to the line splitting of resonance caused by the polar non-uniformity (*Nickolaenko and Hayakawa*, 2002). Most probably, it was conditioned by spectral instability of data sample 5 minutes long or by an interplay of the polar non–uniformity and a special positioning of the field sources (*Rabinowicz,* personal communication).

2.4. The Impact of Extra-Solar Gamma Ray Busts

A couple of intense γ-ray bursts from the exotic neutron stars were detected on the Earth. An attempt was undertaken to find out any changes in the ELF electromagnetic records, which could be associated with modification of the lower ionosphere by the cosmic 1998 γ-ray burst. No distinct signatures of this γ-ray burst were found at ELF (Price and Mushtak, 2001), whereas records of the VLF transmissions revealed significant ionospheric reduction associated with this burst (*Inan et al*., 1999; *Dowden et al*., 1999; *Tanaka et al*., 2008).

Here we deal with a recent huge γ-ray burst from Sagittarius (SGR) 1806-20 that occurred on December 27, 2004, and we model the possible impact of the magnetar on SR. The information on this γ-ray burst event might be found in the literature (e. g., *Terasawa et al*., 2005), but we mention only the necessary parts of it. The center of illuminated zone was found at 146°W and 20°S in the middle of Pacific Ocean, and was positioned at a distance of ~450km from the center of the dayside hemisphere. The intensity was larger by two orders of magnitude than that of SGR+14 in 1998.

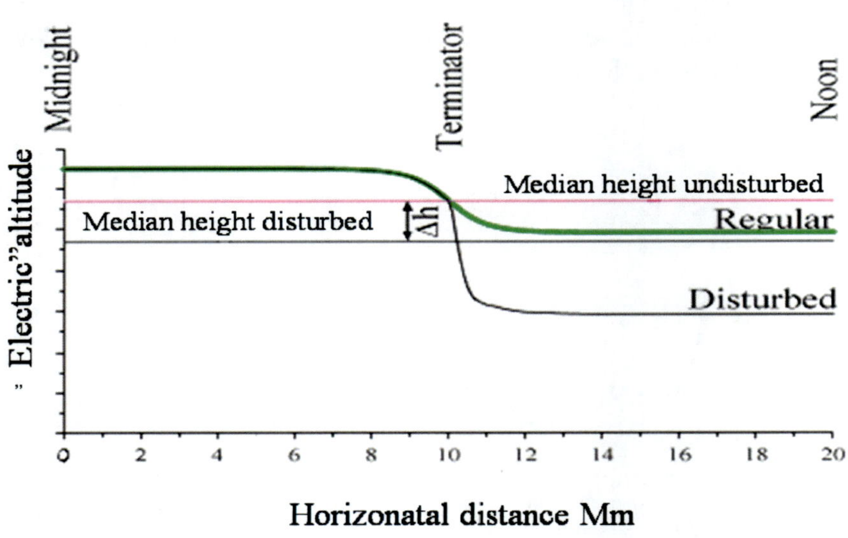

Figure 10. Height of the Earth–ionosphere cavity showing its day–night asymmetry.

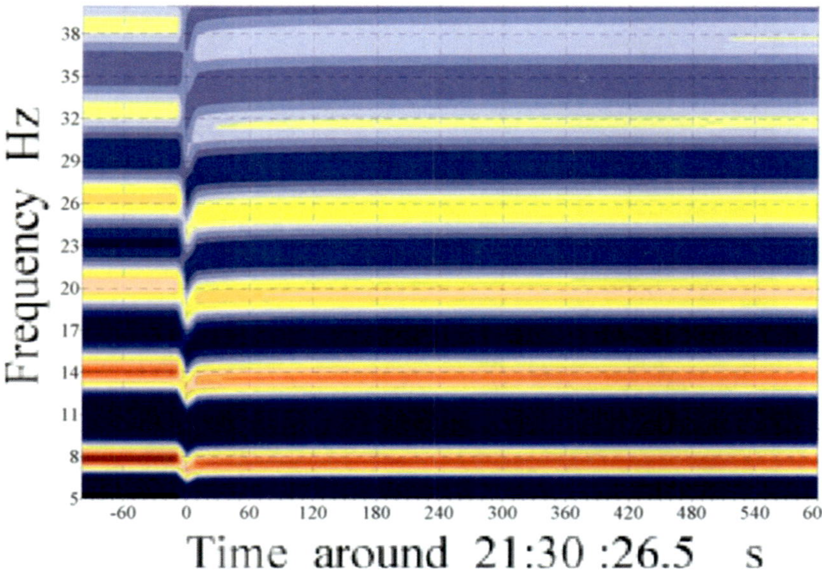

Figure 11. Expected dynamics of SR spectra modifications during the γ – ray burst of Dec. 27, 2004. The time t = 0 corresponds to the γ – burst onset. Spatial distribution of the global lightning activity corresponding to OTD data for 21.5 hr UT.

Figure 10 illustrates schematically the reduction of the dayside ionosphere caused by the γ-ray burst. The abscissa depicts the distance in Mm along the great circle path. The center of the night hemisphere is placed at the origin and the day–night interface (or terminator) is 10Mm from this point. The ordinate shows the "knee height" of the model by *Williams et al.* (2006). The green curve is the regular cavity, and the disturbed profile is seen for the dayside hemisphere. The details are given in Nickolaenko and Hayakawa (2010), but by using the temporal change in ionohpheric height and by assuming the spatial distribution of global lightning activity corresponding to the OTD data for UT=21.5 hr, Figure 11 is the theoretically expected spectra of SR modifications during the γ-ray burst. We observe the distinct changes in the SR spectra, which occur at all the modes simultaneously. We intend to compare these theoretical expectations with the ELF SR observation at Moshiri.

3. SR and Global Temperature

3.1. The North-South Migration of Thunderstorm Activity

The long-term monitoring of SR parameters (peak amplitudes and peak frequencies) provided information on seasonal motion of lightning activity (*Sátori*, 2003; *Sátori et al.*, 2003). Figure 12 illustrates such diurnal/seasonal variations of the first SR frequency. Each pattern arises from the daily motion of thunderstorms around the globe. When the motion takes place along the same parallel, the traces repeat day after day, which is especially obvious in the summer and winter periods (the left plots in Figure 12). At these times, thunderstorms are adjacent to the northern or southern routes.

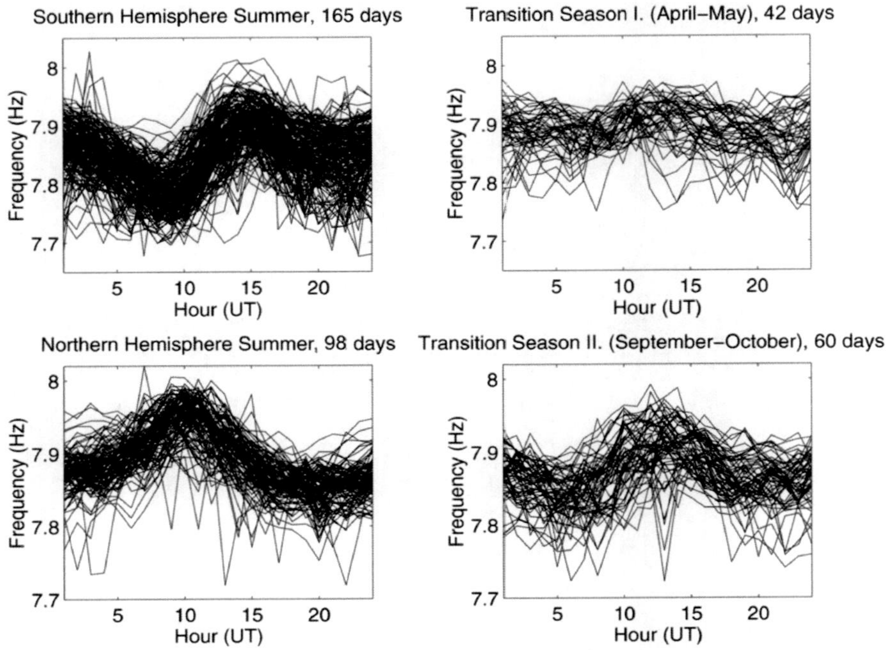

Figure 12. Variations in the diurnal patterns of the first mode frequency caused by the north – south migration of the global thunderstorm activity (adapted from *Sátori* (2003)).

The duration of southern and northern summer periods is 165 and 98 days correspondingly. Intermediate seasons last for 42 (spring) and 60 days (autumn). Hence, the global electric activity rapidly crosses the equator and pauses at the Tropic of Cancer (boreal summer) or at the Tropic of Capricorn (Australian summer), and this behavior is clearly seen in the patterns of resonance frequencies recorded at the middle latitude observatory Nagycenk (47°N, 17°E). The data are consistent with OTD data.

3.2. SR Intensity and Land Temperature

The connection between SR intensity and the planetary tropical temperature anomaly was demonstrated by *Williams* (1992) based on the electromagnetic data collected by C. Polk in 1970s. Modern registrations confirm his result and also allow for a more detailed comparison of resonance intensity with the global temperature.

3.2.1. Hungarian global thermometer

The upper panel in Figure 13 depicts the diurnal/seasonal pattern of the E_Z component. The records were performed at Nagycenk (47° N, 17° E). The lower plots illustrate seasonal variations of the resonance intensity, variations of the land surface temperature in the tropical belt, and the average continental temperature of Northern hemisphere (the circles mark the temperature of North America, stars correspond to Europe, and crosses denote the temperature of Asia). The resonance data are in a good agreement with temperature, especially, when temperatures of continents and of the tropics are combined.

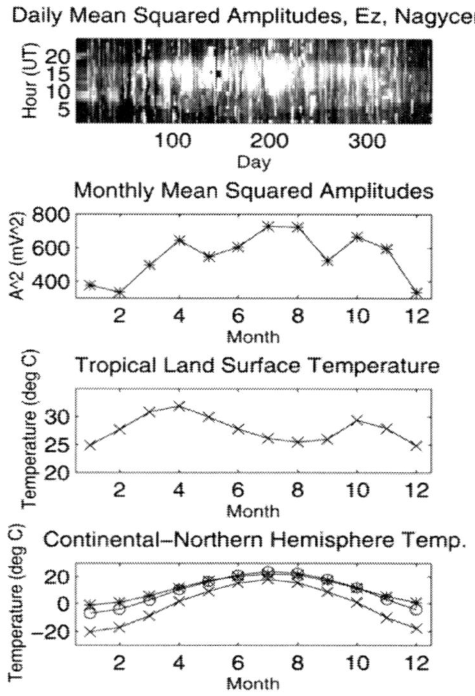

Figure 13. The annual variations of daily patterns of the cumulative SR intensity recorded at Nagycenk in 1995 (top panel). The monthly mean of resonance intensity (second panel); the tropical land surface temperature (third panel); The temperature of continents in the Northern hemisphere {o – North America, * - Europe, x – Asia} (bottom panel). (Adapted from *Sátori et al.*, 2003).

3.2.2. Japanese global thermometer

The links were studied of SR intensity with the global land temperature based on the electromagnetic records at Moshiri observatory, Japan (44°N, 142°E) from November 1998 to May 2002 (*Sekiguchi et al.*, 2004, 2006; *Hayakawa et al.*, 2006). The data are shown in Figure 14, where the upper panel depicts seasonal variations of cumulative SR intensity in the north–south horizontal magnetic field component. The data are given in dB relative to 1 pT^2. The cumulative intensity is the sum of intensities of resonance oscillations at three resonance modes. The lower panel in Figure 14 shows monthly variations of average ground surface temperature in symmetric latitudinal belts. The upper curve corresponds to the latitude interval [20°S to 20°N]. Below, the plots are shown corresponding to the intervals of ±40°, ±60°, and ±80°. The individual curves do not overlap owing to the land cooling toward the poles.

In addition to an outstanding similarity of temporal variations of resonance intensity and ground surface temperature, Figure 14 presents also the 'recovered' data. The thin line in the upper panel depicts the variations of resonance intensity retrieved from the ground surface temperature in the latitude interval ±60°. In contrast, the thin line in the lower panel demonstrates variations of the tropical [20°S – 20°N] temperature recalculated from the SR recorded at Moshiri. There exists a nonlinear connection between the data, which allows us to use the SR intensity as a measure of the global ground surface temperature.

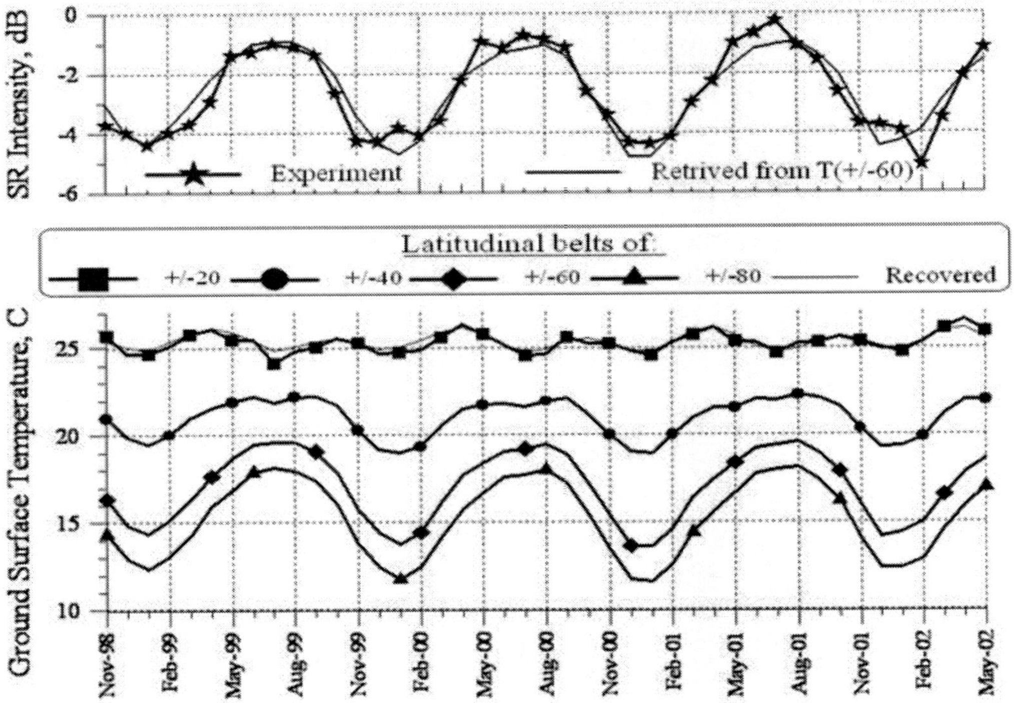

Figure 14. The monthly variations of SR intensity and of the ground surface temperature in different latitudinal intervals (after *Sekiguchi et al.*, 2004).

4. New Subjects for SR Studies

4.1. SR on Other Planets: TITAN (Cassini Mission / Huygens Probe)

Electromagnetic experiments were planned in the framework of the Cassini mission with the Huygens landing probe (by European Space Agency (ESA)). Since the electric activity could be expected in the atmosphere of the Titan Saturnian's moon, preliminary computations were made of possible SR spectra there (see e.g. *Nickolaenko et al.*, 2003). Figure 15 shows the optimistic model spectra of SR on Titan. The steepest vertical conductivity profile was used, and the spectra of a point source were computed for different source–observer angular distances (θ in the figure). The spatial distribution of global resonance oscillations is similar at any planet: the regular modal structure is modified by the interference of the direct and antipodal waves. Properties of the ionospheric boundaries determine the losses in the cavity. Therefore, a general structure is present in different models, however, the resonance pattern may become more or less pronounced depending on the particular ionosphere profile.

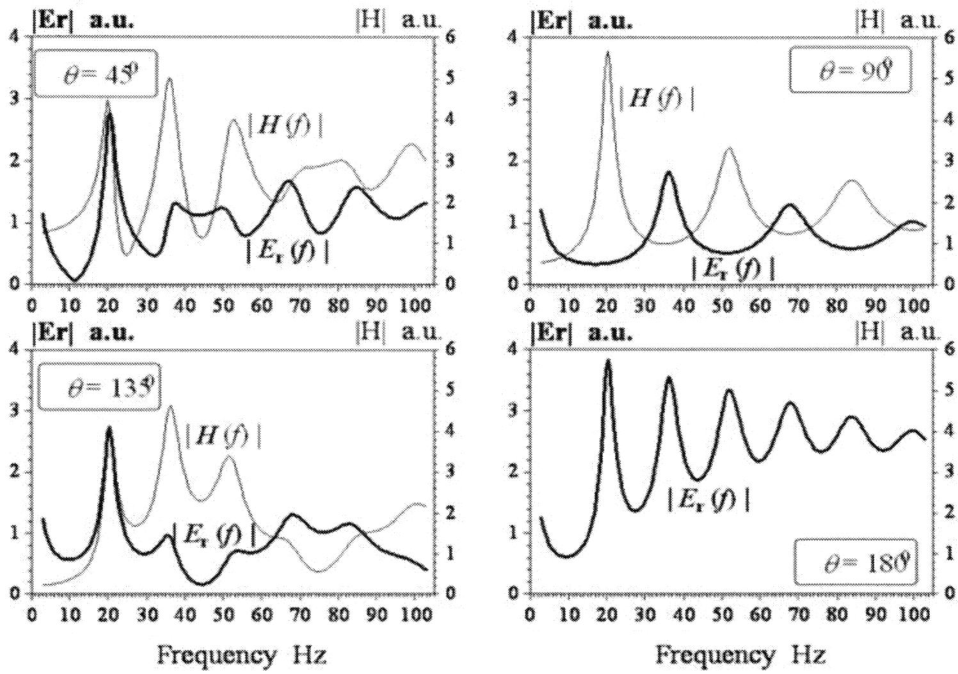

Figure 15. The SR power spectra computed for Titan in the model with optimistic reflection properties of the ionosphere (after *Nickolaenko et al.*, 2003).

4.2. New SR Observatory at Spitzbergen

A new SR observatory started working. It belongs to Poland and is placed at polar latitudes in Horzund, Spitzbergen (77ºN, 15.5ºE) (see *Neska and Sátori*, 2006). Three field components are continuously recorded at the place. The electric ball antenna (*Ogawa et al.*, 1966) and two identical induction coils are used for the detection of two horizontal magnetic fields.

Three panels in Figure 16 illustrate an example of their experimental results of high quality. The upper panel depicts the power spectrum obtained as the Fourier transform of a long signal record. The abscissa depicts the frequency ranging from 3 to 33 Hz, and the ordinate shows the field intensity in the logarithmic scale. The SR pattern outline is evident, and the 'diffusion' of the plot is conditioned by the properties of Fourier transform (the localization principle).

Two lower plots in Figure 16 illustrate the monthly averaged diurnal variations of resonance intensity in the orthogonal magnetic field components recorded simultaneously at Spitzbergen and Rhode Island (47°N, 71°W, USA) in September, 2004. The UT is shown on the abscissa in hours. The intensity variations are shown by the thin (Rhode Island) and the thick (Spitzbergen) lines. The curves reflect the daily motion of thunderstorms around the globe, and we see their close relevance to each other. The peak around 12 hr UT is observed in the lower left plots, which might be associated with the monsoon activity in Himalayas (see below).

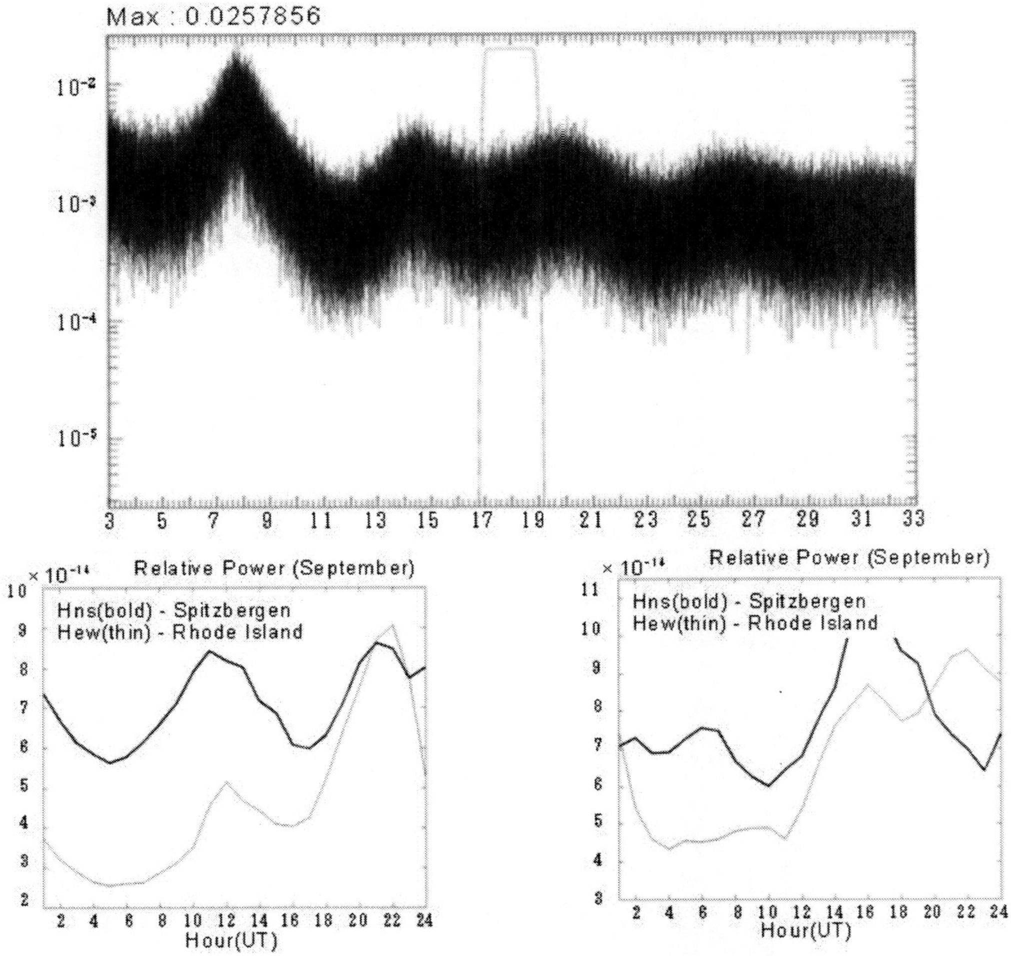

Figure 16. Diurnal variations at Rhode Island and Spitzbergen (adapted from *Neska and Sátori*, 2006).

4.3. Sprites and ELF Transients

Since the discovery of sprites by *Franz et al.* (1990), there has been made a significant progress in the studies of TLEs both from the observational and theoretical points of views. Please refer to a monograph by *Fullekrug et al.* (Eds.) (2006) on the reviews on the observation and theory of TLEs, and also look at several papers in the special issue on SR edited by *Pasko (Ed.)* in Radio Science (2007).

Here we would like to show some of our own recent publications on sprites and ELF transients to give hints on a few essential unsolved problems of TLEs. The most mysterious is the explanation of different shapes of sprites (sprite morphology). Here we want to emphasize the importance of Japanese winter sprites on the Japan Sea side. Winter sprites in the Hokuriku area are found to exhibit a very simple structure, mainly columns (e. g., *Hayakawa et al.*, 2004), which is very much different from the presence of a variety of complex shapes (carrots etc.) for the continental sprites (e. g., *Lyons*, 2006). Figure 17 indicates the percentage occurrence of different sprite structures for the Hokuriku area and Pacific Ocean, which means that winter sprites in the Pacific Ocean is closer to the summer continental ones. This difference in sprite morphology was tried to be interpreted in terms of the difference in spatial structure (especially charge height) of parent lightning discharges (*Asano et al.*, 2008; *Myokei et al.*, 2009). The next subject is the time delay of a sprite behind its parent lightning discharge. Based on the simultaneous observation at VHF and at ELF, *Matsudo et al.* (2009) have compared the time delay of a sprite behind its parent lightning in Figure 18. As the conclusion, the mean time delay is the longest (~90ms) for winter sprites in the Hokuriku area (a) and it is in a medium range for the corresponding winter sprites in the Pacific Ocean (b) as seen in Figure 18. Whereas, the time delay for summer continental sprites seems to be smallest (0-30ms). In order to attempt to explain this difference, *Asano et al.* (2009) have utilized the electromagnetic (EM) computer code in which the quasi-electrostatic (QE) electric field (*Pasko*, 2006) and EM effects are simultaneously included. Then, they have concluded that higher-frequency components (such as M components) in the continuing current might be responsible for long-delay sprites because M components are often involved in the continuing current for the winter lightning in Japan. Finally, the lateral shift is interpreted in terms of a complicated structure of a parent lightning (like a bent lightning discharge) (*Asano et al.*, 2009).

Global mapping of ELF transients has been performed (*Greenberg and Price*, 2004; *Yamashita et al.*, 2009) in order to study the detailed characteristics of huge lightning with both polarities and to find their correlation with sprites.

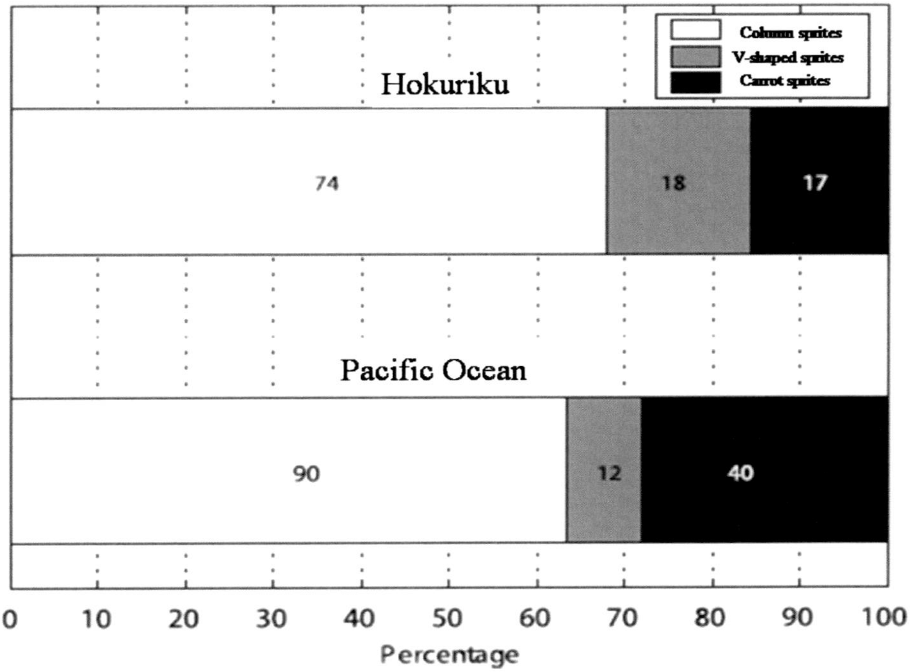

Figure 17. Morphology of optical sprites observed in the Hokuriku (top) and the PacificOcean (bottom). The shapes of sprites are classified into three categories: columns, V-shaped and carrots. Adapted from *Matsudo et al.* (2009).

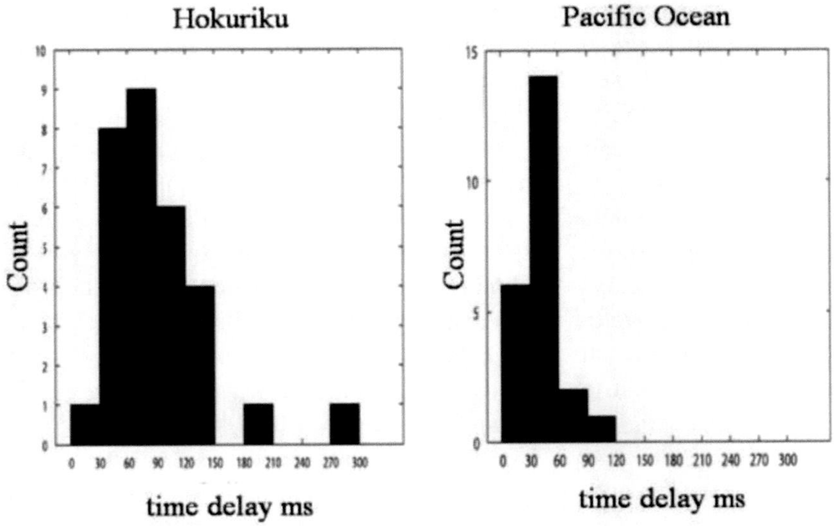

Figure 18. Occurrence histograms of the time delay of a sprite from its causative ELF lightning around the Hokuriku area as observed from Shimizu, Japan (left). The time delay is given with a bin of 30ms. The result for the Pacific Ocean area (right) . Adapted from *Matsudo et al.* (2009).

4.4. Biological Effects of SRs

There are cases of changes in the natural electromagnetic fields linking to any observable changes in higher life form can be found in the scientific literature. Those natural effects include solar activity, geomagnetic activity, cosmic-ray activity, lightning activity etc., and recently *Cherry* (2003) suggested the important role of SRs in the biological effect.

We have studied this subject on the basis of our own ELF observation in Moshiri and the simultaneous observation of human blood pressure, heart rate and depression etc. (*Mitsutake et al.*, 2005). These data are based on the observation during April to July, 2001. As a result, it is found that the blood pressure shows statistically significant group mean differences between normal and enhanced SR days. That is, the group mean of blood pressure is found to show a significantly lower value for enhanced SR days than that for normal SR days. The next physical step is to consider the mechanism for this difference.

5. New Results in the Old Problems

5.1. The Solution of Inverse SR Problem

The integrated SR spectra (so-called 'background' resonance spectra) were used for resolving the inverse problem (*Shvets*, 2001) in an attempt to evaluate the position (or map) of the global thunderstorms. Recently, the problem was treated by using the FDTD technique (*Ando and Hayakawa, 2004*). Figure 19 shows a sample of results of our numerical tests of the algorithms applied. The left plot shows the postulated resonance spectra: the initial (smooth) one and that containing the 3% 'noise'. The right plot depicts the postulated distance distribution of field sources (bars) and those inferred from the solution of inverse SR problem. Generally, the source distribution is recovered fairly well from the resonance record. The approach was applied to the real records performed at Moshiri observatory (Japan), and the diurnal dynamic spectrum was obtained of the source distance and intensity from March to August (not shown). These results were in accord with the previous studies and climatological data (see *Ando et al.*, 2005 for details), and it is suggested that this kind of inverse problem would be useful for obtaining the global lightning mapping.

The first application of the above-mentioned inverse problem to the real SR data actually and simultaneously observed at these stations (Moshiri, Rhode Island and Lekhta stations) has been done by *Shvets et al.* (2009). The inversion technique itself is so time-consuming and laborious that we have used only two day data. We have employed a two-step approach for reconstruction of the global lightning activity by a network of SR stations. The first step consists of inversion of SR power spectra measured in a wide frequency range at each station into one-dimensional source distribution; distance profile (as mentioned above), in which we use a spherically symmetrical model of the Earth-ionosphere cavity. The second step is the tomographic reconstruction of a desired two-dimensional source distribution from a set of projections, which are based on source distance profiles obtained simultaneously at every station of the network.

The details of the inversion technique have been found in the paper by *Shvets et al.* (2009), and we show only one example of the reconstructed global distribution of lightning discharges in Figure 20 on a particular day of January 9, 2000. The UT is 8-9h for the top panel, UT=15-16h for the middle and the bottom is for UT=18-19h, which are known to correspond to the main time interval of active lightning activity in South-east Asia, Africa and America, respectively. The color contours indicate the distribution of lightning activity. In each panel the main lightning-active area is clearly reconstructed, so that this inversion technique would be a unique method to obtain the global lightning map as a snapshot. We do hope that the inversion should be performed for a much longer data set, and this work is underway.

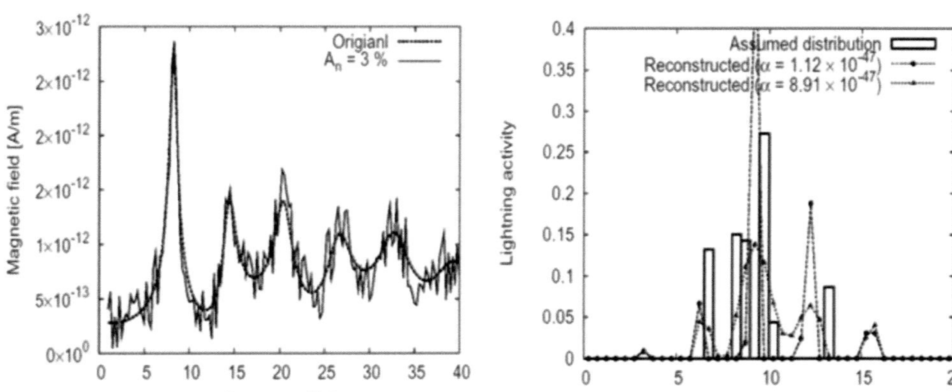

Figure 19. The numerical experiment with solution of the inverse problem. Left – the original and 'noisy' spectra (3% noise) and right - assumed (bars) and its reconstructed (lines) source distance distributions with different values of Tikhonov regularization parameter.

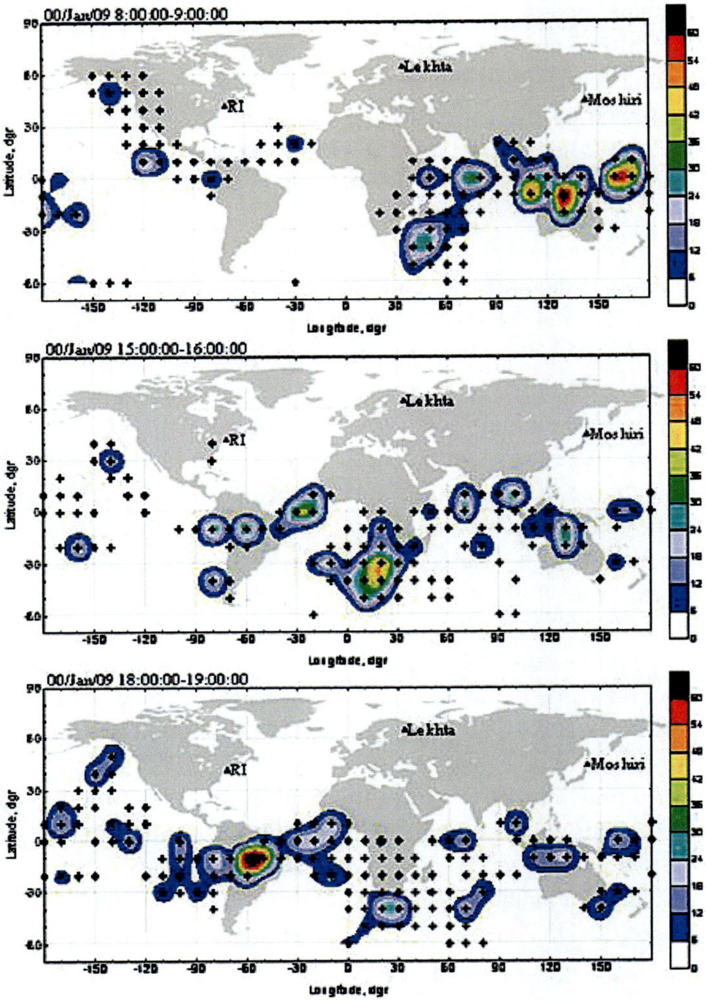

Figure 20. Examples of lightning intensity maps reconstructed from three-station SR measurements for the time intervals during a day of January 09, 2000, corresponding to dominant lightning- active centers (from the top to the bottom panel) in Asia, Africa, and South America. Positions of the observatories are marked by small triangles with station names (adapted from *Shvets et al.* (2009)).

5.2. 'Terminator Effect'

Our attention was devoted to the 'terminator effect' observed in the long–term SR records (*Melnikov et al.*, 2004). We reproduce in Figure 21 the lens-like diurnal/seasonal pattern of the SR intensity observed at Mitspe Ramon (31°N, 35°E) in 1992 – 2002 and Nagycenk (47°N, 17°E) in 1994 – 2001. The middle panel shows the seasonal change in the sunrise and sunset times at these two observatories. The variations observed were attributed to vatiations in the local ionosphere height above the observatories (*Melnikov et al.*, 2004). However, the model computations of the terminator effect published in 1980s predicted the minute modifications of the field amplitude.

Weakness of the terminator impact on the SR was established in the theoretical works since 1960 (see *Nickolaenko and Hayakawa*, 2002 for reference). The similar results were obtained in recent model studies. Pechnoy and Price (2004) used the 2D (two dimensional) telegraph equations, which suggest the 2D transmission line model of the spherical Earth–ionosphere cavity. The parameters of transmission line were found from the partially uniform knee model of the ionosphere, which explores the abrupt day–night transition. The experimental and model data are compared in Figure 22 for two field sites: Rhode Island and Mitspe Ramon. The upper plots show the experimental diurnal/seasonal patterns of the first mode intensity. The months are shown on the abscissa, and local time (LT) is depicted on the ordinate in hr. Two lens-like patterns are seen in the experimental data. The middle plots in Figure 22 indicate the model data computed for the uniform Earth–ionosphere cavity. The lower plots show the changes in the presence of the day-night asymmetry.

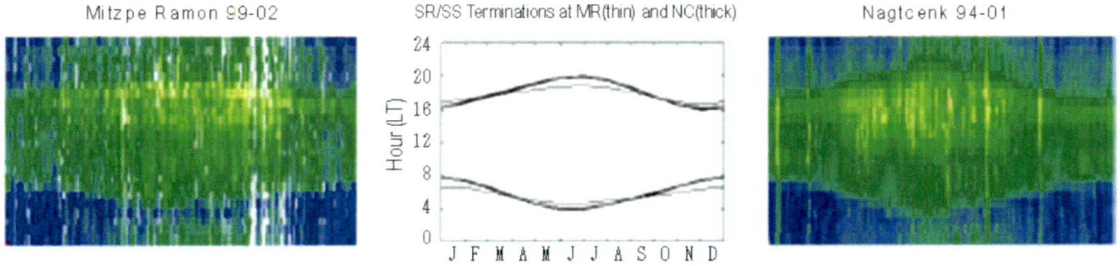

Figure 21. The third SR mode E_Z diurnal/seasonal variations at Mitspe Ramon (left panel) and Nagycenk (right) with the computed terminator times for these two stations (middle panel) (adapted from *Melnikov et al.* (2004)).

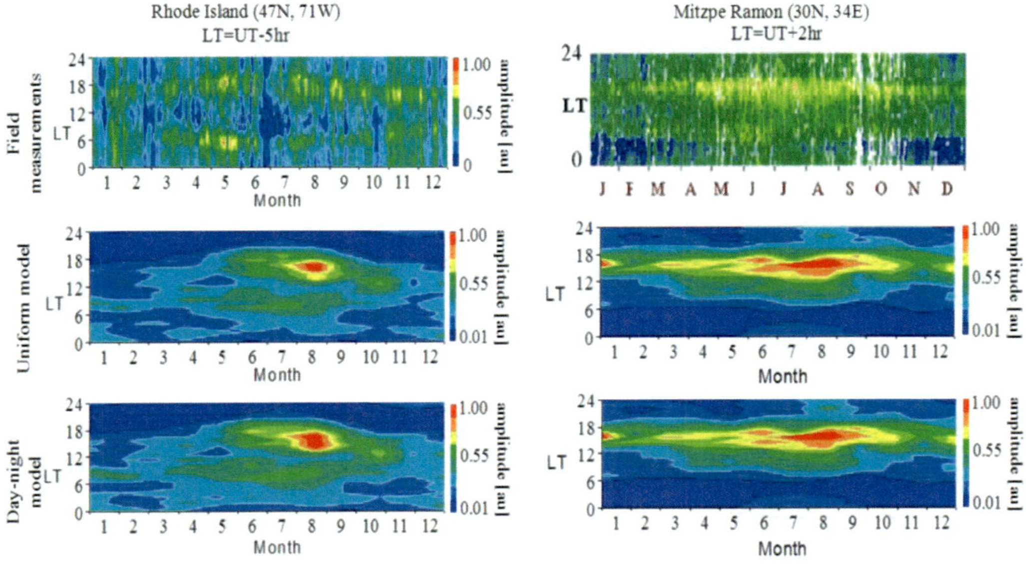

Figure 22. Experiment versus model data. The model gives approximately 5% terminator effect. Shown are the diurnal/seasonal variations in the $|E_Z|$ component at the first SR mode at Mitzpe Ramon and Rhode Island (after *Pechnoy and Price* (2004)).

The model data of Figure 22 exploit the space–time distribution of thunderstorm activity obtained by the OTD. Figure 22 demonstrates general similarity of the model maps to those obtained experimentally. Our computations in the uniform and non-uniform cavities indicate that the patterns must be attributed to the source dynamics. The day–night transition itself provides a secondary effect, which does not exceed the 7% amplitude modulation that agrees with earlier publications in 1980s.

The OTD data allow for obtaining more or less realistic diurnal/seasonal dynamics at an arbitrary observatory within the uniform Earth–ionosphere cavity model. We show, in Figure 23, the variations computed for different observatories all over the world. Each plot here combines twelve separate strips which correspond to a particular month. The strips comprise the 24 hr diurnal variation of resonance amplitude, and thus the plots depict 12 successions of diurnal patterns varying from month to month. The lower plot in the figure shows changes of the global thunderstorm activity itself.

As we see from Figure 23, the cumulative intensity of global resonance reflects changes in the thunderstorm activity worldwide. The deviations are present at different sites caused by the particular source–observer geometry. However, the 'start' and the 'end' of the SR oscillations occur simultaneously all over the globe relevant to the abrupt increase in activity around 8–10 hr UT. A relevant threshold is shown in the plots by the horizontal lines, and we observe that the cease of activity also tends to occur simultaneously.

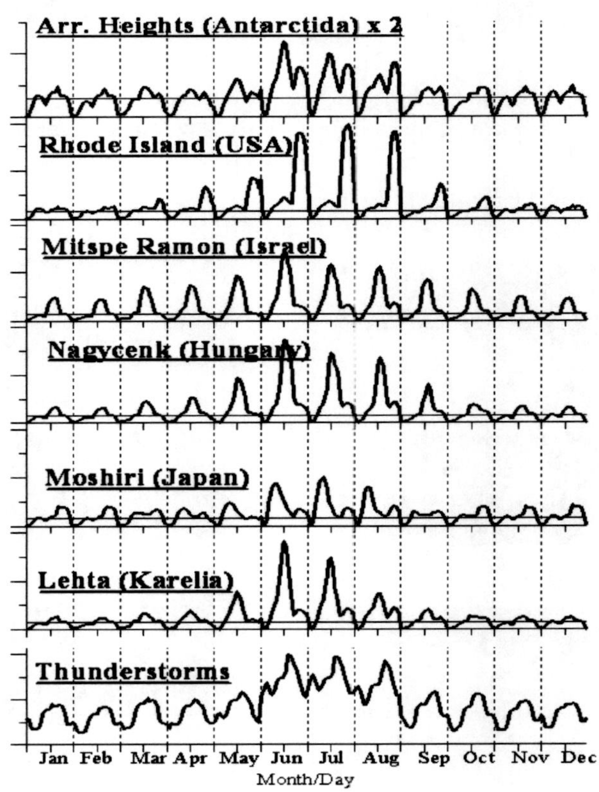

Figure 23. Monthly variations of the daily patterns in the uniform Earth –ionosphere cavity model computed from OTD data.

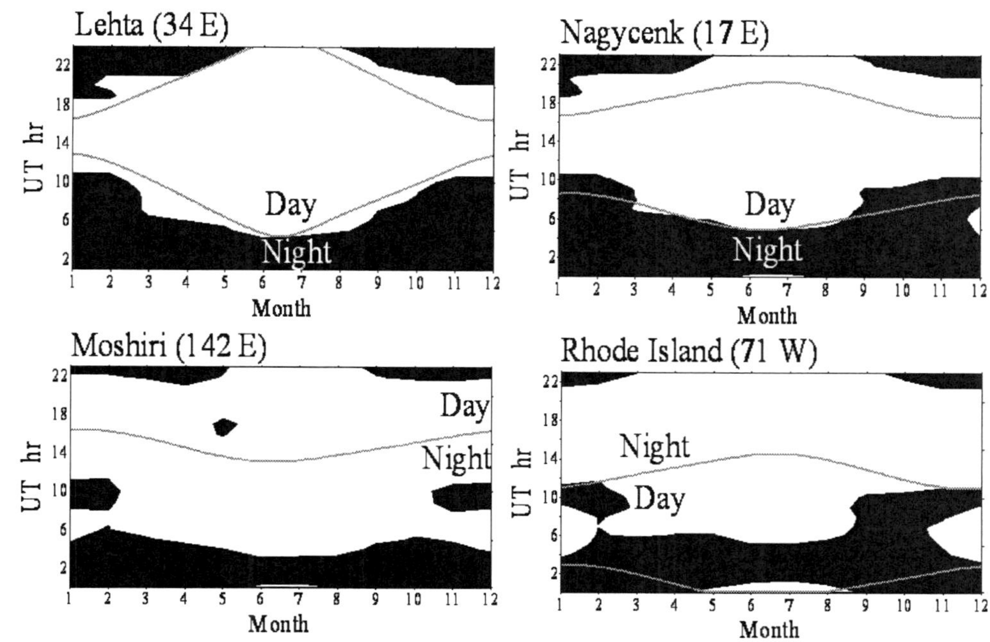

Figure 24. Maps showing the SR patterns and local sunrise/sunset times (given in orange lines). Day is given in white, and Night, in black. The correspondence is good for Europe, but unsatisfactory at other observatories.

The property becomes especially clear when we plot the data in the same fashion as in Figures 22 and 24. The month is shown on the abscissa, and the time UT is shown on the ordinate in hr. The resonance intensity is shown by inking. When the cumulative intensity exceeds the threshold of Figure 23, the plot is white. When it is below the threshold, the plot is black. Four longitudinally separated observatories were shown: Lehta, Nagycenk, Moshiri, and Rhode Island. The relevant times of local sunset and sunset are also shown in the maps by lines. Figure 24 demonstrates the global property of SR: the oscillations globally increase and decrease in accordance with the start/stop of the planetary thunderstorm activity. The local properties (not shown in the figure for simplicity) are also present in the data, including the properties that reflect the source proximity to the observer.

5.3. The Location of Q-Bursts and the Relevant Accuracy

An interest reappeared toward the Q–bursts described by T. Ogawa group in 1960s (*Ogawa et al.*, 1966, 1967). The method was suggested by D. Jones for the source distance finding from the resonance pattern (*Jones*, 1970; *Jones and Kemp*, 1970). The transient ELF pulses might be associated with the TLEs in the mesosphere (red sprites). A coordinated measurement campaign was conducted that included the ground-based detection of TLE-producing intense lightning during the MEIDEX mission on board the space shuttle Columbia (*Price et al.*, 2004).

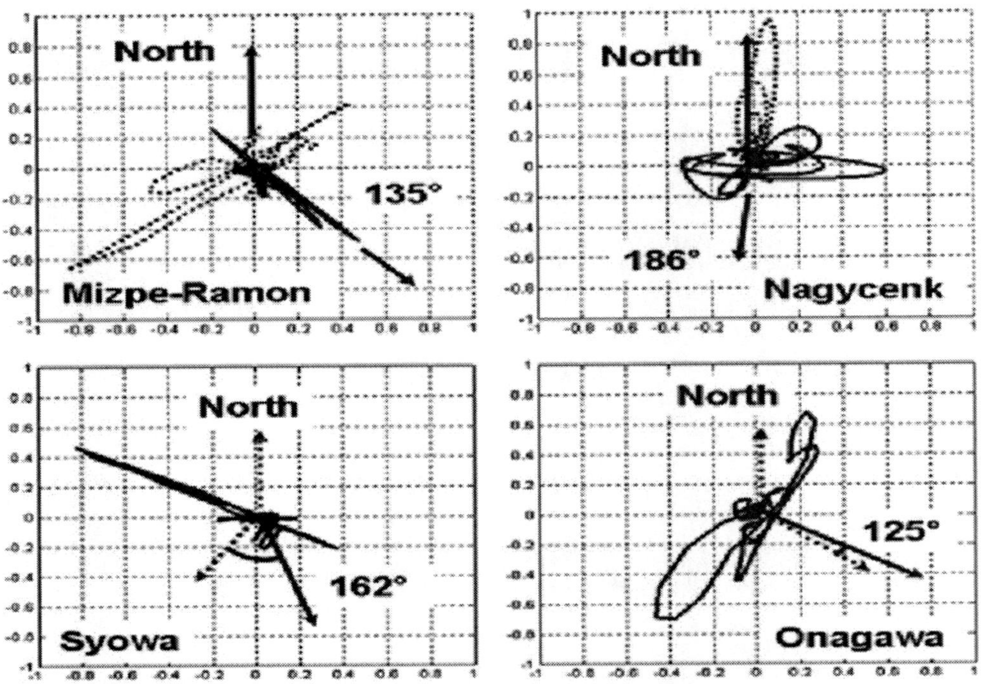

Figure 25. The global lightning location based on the electromagnetic signatures in the SR band (adapted from *Price et al.*, 2004).

Figure 25 shows the Lissajous plots of the horizontal magnetic field component and of the Poynting vector of a particular ELF transient acquired at the observatories in different parts of the world. The solid line depicts variations of the magnetic field, and the dotted line indicates the evolution of Poynitng vector. The directions to the causative strokes are also shown. We must separately mention here an original technique of the direction finding suggested and used by *Sato and Fukunishi* (2003). It exploits the waveforms of the Q–burst magnetic fields only, but allows for the unambiguous location of the parent stroke.

These works were conducted for studying the global sprite occurrence by detecting the Q–bursts originating from the strokes. When performing such a task, we must be cautious, because the direction finding based on the SR field components (instead of those of Poynting vector) might be associated with serious uncertainties. Figure 26 explains the worries.

When one detects a Q–burst, its pulsed signal arrives concurrently with the background resonance signal. We assume that the signal to noise ratio (SNR) is $R = 5$, which means that the burst amplitude is five times the RMS background signal amplitude. Depending on the source azimuth, the SNR varies being always smaller than 5. The most accurate source bearing is obtained within the zone of equal signal. In a case when the wave arrival angle approaches one of the cardinal directions, the pulsed signal is hidden in the background noise of one of the channels. We assume that the source bearing is 'accurately recovered' when the relevant ratio exceeds unity. Otherwise, the arrival angle cannot be established accurately, and we regard the relevant sector as the 'sector of unconfidence'. Figure 26 depicts such sectors. Each of them is centered at a cardinal direction, as the cores of magnetic antennas are aligned with these directions. The sectors are 24° wide, and put 'out of game' 26.6% of the globe. When the observations are conducted at two sites, and the source coordinates are found from the triangulation, individual 'sectors of unconfidence' are summed up, since the source position in any of the 'unconfidence sectors' for any site prevents the secure location.

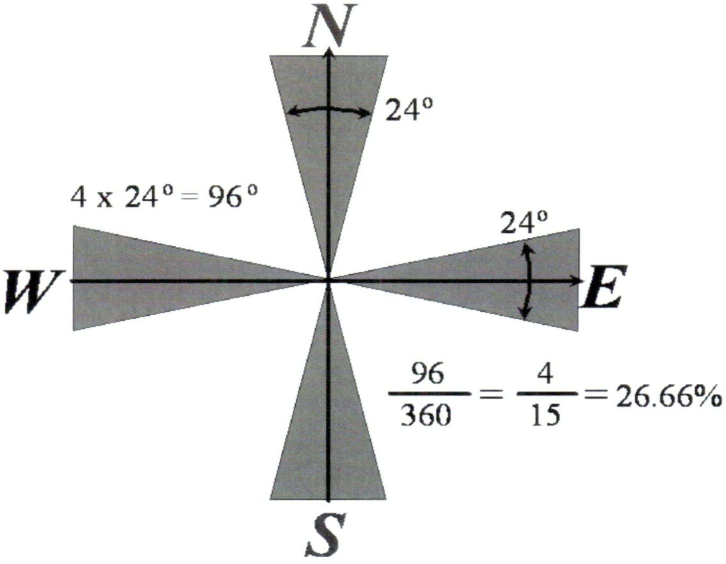

Figure 26. The cross–like zone of tentative direction finding for the burst/background amplitude ratio R = 5, and the uncertainty ratio of 1.

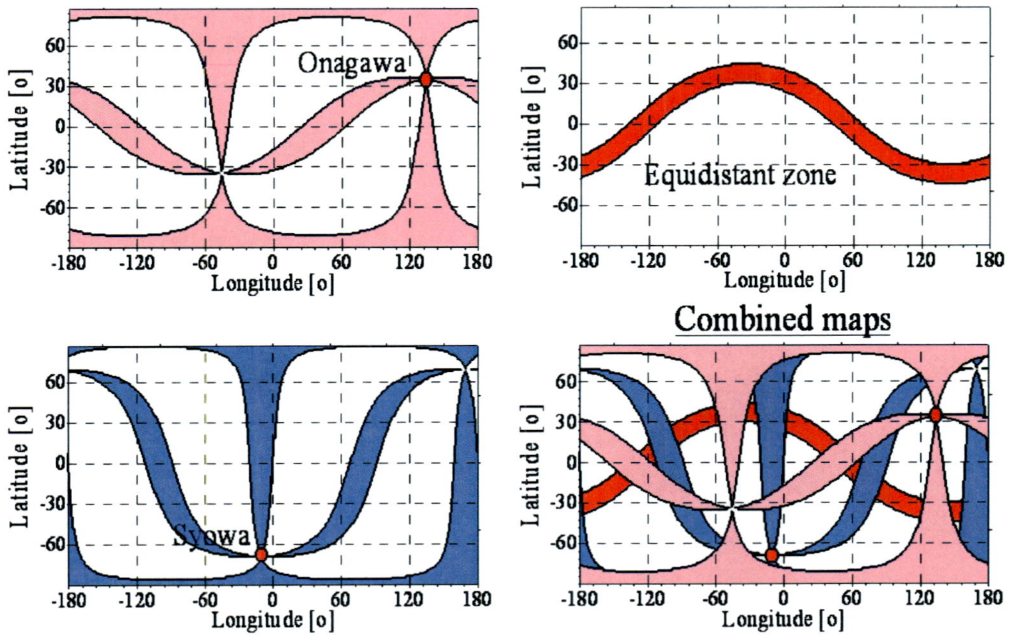

Figure 27. Zones of unconfident source location for the Syowa and Onagawa observatories on the assumption that the antennas are oriented along geographical cardinal directions.

Two left panels in Figure 27 depict the global maps of 'unconfidence sectors' computed for the Onagawa (38° N, 142° E) and Syowa (69°S, 40° E) observatories used in *Sato and Fukunishi* (2003) (We ignore the magnetic declination here). These maps must be combined with the equidistant unconfidence zone, and the result is shown in the left lower frame. More than 50% of the territory falls into the unconfident location zone. Therefore, the position of the Q–burst parent strokes must be sought with great caution. In particular, the Poynting vector technique should be preferred as the one that effectively increases the SNR by a factor reaching an order of magnitude (*Nickolaenko and Hayakawa*, 2002).

5.4. A Comparison of Waveforms of Q-Bursts (Observation versus Theory)

The sampling frequency of the former ELF observation has been less than 100Hz or so, so that a comparison of waveforms of Q-bursts between the experiment and theory has been impossible. When we have a wide-band observation, it may be possible to perform a comparison of Q-burst waveform between the experiment and theory.

Experimental waveforms of natural ELF transient signals (Q-bursts) are compared with the computations based on the analytical time domain solution (*Nickolaenko et al.*, 2008).The computations are performed for the uniform (without terminator effects) Earth-ionosphere cavity model with a linear frequency dependence of the ELF propagation constant. The vertical electric field for ELF observation was recorded in the frequency band from 1Hz to 11 kHz by means of a ball antenna in fair weather conditions at Kochi (133.3°N, 133.4°E) during the period of 2003-2004 (*Ogawa and Komatsu*, 2007). An example of comparison is shown in Figure 17, in which the thin line refers to the experiment, and the thick line, the theoretical waveform for the source-observer distance of 17.4Mm。 Waveforms of discrete pulses in Figure 28 are compared, which indicates a high similarity. This confirms the validity of the model and the consistency of the source-observer distance found experimentally, directly in the time domain.

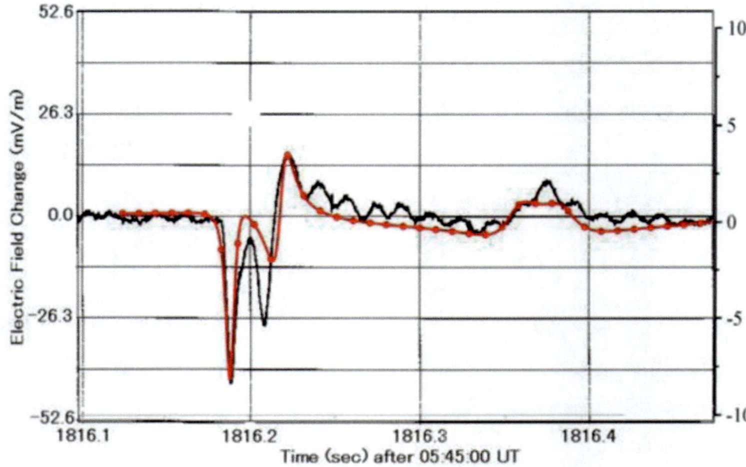

Figure 28. Comparison of experimental (smooth black curve) and model (red curve with dots) waveforms.

5.5. Parameters of Thunderstorms Inferred from the OTD Data

The optical observations of lightning flashes from space presented a basis of data used in modeling the SR. We demonstrate general properties of the global lightning activity that follows from the OTD records and compare them with classical meteorological data (*Hayakawa et al.*, 2005b; *Nickoklaenko et al.*, 2006b). The upper map in Figure 29 demonstrates the global distribution of lightning flashes following from OTD records. The stoke distribution repeats the continental outline.

The lower map of Figure 29 shows dynamics of lightning activity in the time–longitude coordinates. We observe here daily changes of the flash number (the standard longitudinal interval was used of 2.5°). Three tilted lines show the drift of the fixed local times of 15, 16, and 18 hr. The maximum of activity occurs around 16 hr LT, and it it substantially reduced over the oceans. We may note that Figure 29 shows how thunderstorms 'step over' the oceans from one continent to another during the course of daily motion.

There is a compact thunderstorm center in Himalayas in Figure 29 with the peak activity occurring around 10–12 hr UT. This center might be responsible for the 12 hr peaks detected in the H_{NS} component at Spitzbergen and in H_{EW} record at Rhode Island, as seen in Figure 16.

We show the drift and longitudinal redistribution of thunderstorm activity in Figure 30. The collection of individual profiles in the upper frame depicts the spatial drift of the flash rate, and the westward motion of activity. The lower diagram in this figure presents an average longitudinal distribution of activity, which contains three distinct global thunderstorm centers at Africa, America, and Asia.

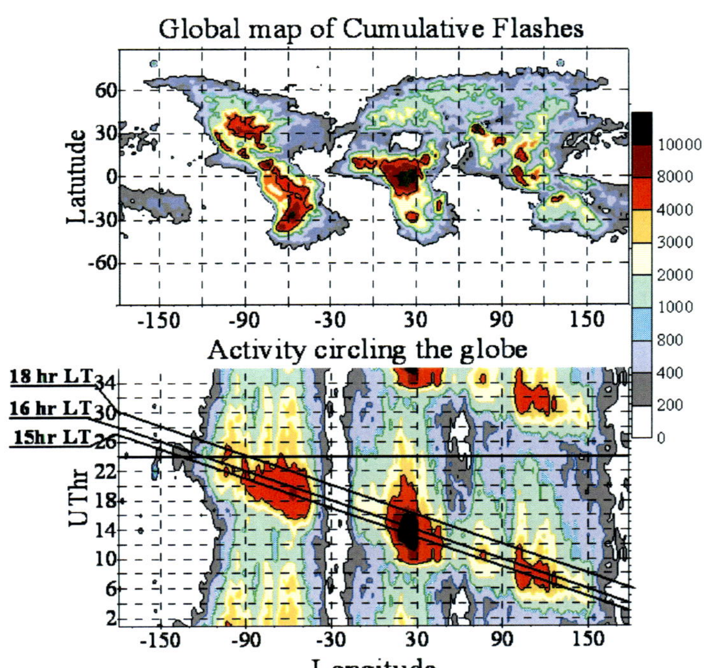

Figure 29. The global lightning distribution and daily motion of lightning activity along the longitude. Three straight lines show the uniform daily motion of the point of 15, 16 and 18 hr LT.

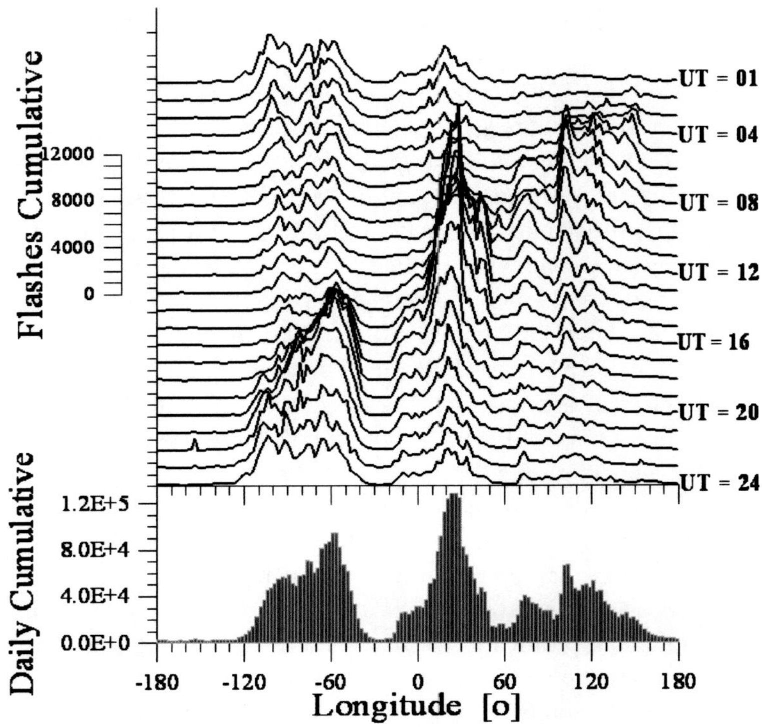

Figure 30. Daily variations in the longitudinal distribution of lightning strokes.

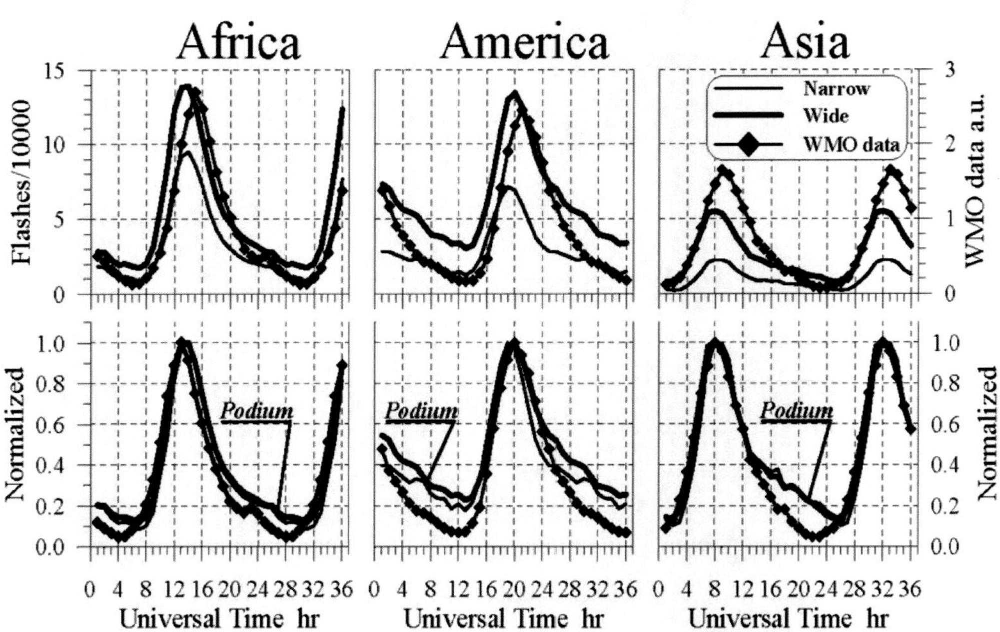

Figure 31. Diurnal variations of intensity in the major global thunderstorm centers.

The temporal variations are depicted in Figure 31 showing significant activities at three centers (Africa, America, and Asia). We compare the modern optical data with the classical WMO records marked by diamonds. The OTD data were integrated in the narrow and wide longitudinal intervals around the centers (see Figure 30). We can notice an excellent reciprocity of the modern to classical data. However, two features might be noted. Firstly, the intensity of Asian thunderstorms seems to increase in comparison with the WMO data. Secondly, the 'podium' or the nocturnal activity at every center appeared to be higher in the updated measurements from space.

The OTD data correspond to the classical WMO records, and they also agree with experimental SR records. Therefore, the optical data become a useful tool for the realistic modeling of the processes in the Earth–ionosphere cavity when observed at different sites.

6. Conclusion

In this review we have intended to survey the recent activities on ELF SRs and transients mainly during the last decade, though it is based on our own choice and preference. In the section of effects connected with the non-uniformities in the lower ionosphere, we have included the seismic effects, the proof of the SR line splitting, and the impacts of solar effect (solar proton events) and the extra-solar effect (γ-ray bursts from exotic neutron stars). Then, Section 3 has dealt with the relationships between SRs and global temperature, in which positive correlation between the two is reported from the Hungarian and Japanese data. Section 4 is concerned with the new topics related with ELF; SRs on other planets, a new SR observatory, sprites and ELF transients, biological effects of SRs. The last section 5 deals with the new results for the old SR problems, which includes (1) inverse problems of SRs, terminator effects, location accuracy of Q-bursts, a comparison of waveforms of Q-bursts (experiment and theory) and parameters of thunderstorms inferred from the OTD data.

Acknowledgments

The work by A.P.N. was supported in part by the Science–Technical Center of the Ukraine (STCU) under the grant #2070. This work is also supported by NICT (National Institute of Information and Communications Technology) and Sekisui Chemical Foundation, to which we are grateful.

References

Ando, Y. & M. Hayakawa, 2-D finite difference analyses of Schumann resonance and identification of lightning distribution, *Inst. Electr. Engrs. Japan, Trans. Fundamentals and Materials, 124,* 1225-1231, 2004.

Ando, Y., M. Hayakawa, A. Shvets, and A. Nickolaenko, Finite difference analyses of Schumann resonance and reconstruction of lightning distribution, *Radio Sci., 40,* No. 2, doi:10.1029/ 2004RS003153, 2005.

Asano, T., M. Hayakawa, M. Cho, and T. Suzuki, Computer simulations on the initiation and morphological difference of Japan winter and summer sprites, *J. Geophys. Res., vol. 113*, A02308, doi:10.1029/2007JA012528, 2008.

Asano, T., T. Suzuki, M. Hayakawa, and M. G. Cho, Three-dimensional EM computer simulation on sprite initiation above a horizontal lightning discharges, *J. Atmos.Solar-terr. Phys., vol.71*, 983-990, 2009.

Basler, M. & C. Wagner, Measurement of the spectrum of radio noise from 50 to 100 c/s, *J. Res. NBS, 64D*, 415-418, 1960.

Basler, M. & C. Wagner, Diurnal power variations of the earth-ionosphere cavity modes and their relationship to worldwide thunderstorm activity, *J. Geophys. Res., 67,* 619-625, 1962.

Boccippio, D. J., E. R. Williams, S. J. Heckman, W. A. Lyons, I. T. Baker, and R. Boldi, Sprotes, ELF transients, and positive ground strokes, *Science, 269*, 1088-1091, 1995.

Cherry, N. J., Human intelligence: The brain, an electrtromagnetic system synchronized by the Schumann resonance signal, *Medical Hypotheses, 60* 843-844, 2003.

Christian, H. J., R.J. Blakeslee, D.J. Boccippio, W.L. Boeck, D.E. Buechler, K.T. Driscoll, S.J. Goodman, J.M. Hall, W.J. Koshak, D.M. Mach, and M.F. Stewart, Global frequency and distribution of lightning as observed from space by the Optical Transient Detector, *J. Geophys. Res., 108(D1),* 4005, doi:10.1029/2002JD002347, 2003.

Dowden, R. L., N. R. Thomson, C. J. Rodger and M. Hayakawa, VLF perturbation caused by Gamma flare, presented at URSI GA (Toronto), G2. P. 2, 1999.

Franz, R. C., R. J. Nemzek, and J. R. Winckler, Television image of a large upward electrical discharge above a thunderstorm system, *Science, 249, 48*, 1990.

Fullekrug, M., E. Mareev, and M. J. Rycroft (Eds.), Sprites, Elves and Intense Lightning Discharges, NATO Science Series, *Springer, 398p*, 2006.

Greenberg, E. & C. Price, A global lightning location algorithm based on the electromagnetic signature in the Schumann resonance band, *J. Geophys. Res., 109*, D21111, doi:10.1029/2004JD004845, 2004.

Hayakawa, M., T. Nakamura, Y. Hobara and E. Williams, Observation of sprites over the Sea of Japan and conditions for lightning-induced sprites in winter, *J. Geophys. Res., vol. 109*A01312, doi:10.1029/2003JA009905, 2004.

Hayakawa, M., K. Ohta, A. P. Nickolaenko and Y. Ando, Anomalous effect in Schumann resonance phenomena observed in Japan, possibly associated with the Chi-chi earthquake in Taiwan, *Ann. Geophysicae, 23*, 1335 – 1346, 2005a.

Hayakawa, M., M. Sekiguchi and A. P. Nickolaenko, Diurnal variation of electric activity of global thunderstorms deduced from OTD data, *J. Atmos. Electr., 25*, 55 – 68, 2005b.

Hayakawa, M., M. Sekiguchi, Y. Hobara, and A. P. Nickolaenko, Intensity of Schumann resonance oscillations and the ground surface temperature, *J. Atmos. Electr., 26*, No.2, 79-93, 2006.

Hayakawa. M., A. P. Nickolaenko, M. Sekiguchi, K. Yamashita, Y. Ida, and M. Yano, Anomalous ELF phenomena in the Schumann resonance band as observed at Moshiri (Japan) in possible association with an earthquake in Taiwan, *Natural Hazards Earth System Sci., 8*, 1309-1316, 2008.

Inan, U. S., N. G. Lehtinen, S. J. Lev-Tov, M. P. Johnson, T. F. Bell, and K. Hurley, Ionization of the lower ionosphere by γ-rays from a magnetar: Detection of a low energy (3-10 keV) component, *Geophys. Res. Lett., 26*, 3357-3360, 1990.

Jones, D. L1., The calculations of the Q-factors and frequencies of Earth-ionosphere cavity resonances for a two-layer ionosphere model, *J. Geophys. Res., 69*, 4037-14041, 1964.

Jones, D.Ll., Propagation of ELF pulses in the Earth-ionosphere cavity and application to slow tail sferics, *Radio Sci., 5,* 1153-1163, 1970.

Jones, D.Ll. and D.T. Kemp. Experimental and theoretical observations of Schumann resonances*, J. Atmos. terr. Phys., 32*, 1095-1108, 1970.

Lyons, W. A., The meteorology of transient luminous events - An introduction and overview, in "Sprites, Elves and Intense Lightning Discharges", Ed. by M. Fullekrug et al., *NATO Sci. Series*, 19-56, Springer, 2006.

Madden, T. & T. Thompson, Low-frequency electromagnetic oscillations of the Earth0ionosphere cavity, *Rev. Geophys., 3(2),* 211, 1965.

Matsudo, Y., T.Suzuki, K.Michimoto, K.Myokei, and M.Hayakawa, Comparison of time delays of sprites induced by winter lightning flashes in the Japan Sea with those in the Pacific Ocean, *J. Atmos. Solar-terr. Phys., 71*, 101-111, 2009.

Melnikov, A., C. Price, G. Sátori, and M. Füllekrug, Influence of the solar terminator passages on Schumann resonance parameters, *J. Atmos. Solar-terr. Phys., 66*, 1187 – 1194, 2004.

Mitsutake, G., K. Otsuka, M. Hayakawa, M. Sekiguchi, G. Cornelissen, and F. Halberg, Does Schumann resonance affect our blood pressure?, *Biomedicine and Pharmacotherapy, 59*, s10-s14, 2005.

Myokei K, Y. Matsudo, T.Asano, T.Suzuki, Y. Hobara, K.Michimoto, and M.Hayakawa, A study of the morphology of winter sprites in the Hokuriku area of Japan in relation to cloud charge height, *J. Atmos. Solar-terr. Phys., 71*, 597–602, 2009.

Neska, M. & G. Sátori, Schumann resonance observation at Polish Polar Station at Spitzbergen as well as in Central Geophysical Observatory in Belsk, Poland, Translation from Polish, Przeglad Geofizyczny, 2006.

Nickolaenko, A.P., ELF radio wave propagation in a locally non-uniform Earth-ionosphere cavity, *Radio Sci., 29*, 1187 – 1199, 1994.

Nickolaenko, A.P. & M. Hayakawa, *Resonance in the Earth-ionosphere Cavity*, Kluwer Academic Publishers, Dordrecht/Boston/London, 2002, 380 pp.

Nickolaenko, A.P., Rabinowicz, L.M., Shvets, A.V. & A.Yu. Schekotov, Detection of Schumann resonance line splitting, *Radiofizika and Electronika, 7, No.3*, 498–508, 2002 (in Russian).

Nickolaenko, A.P., B.P. Besser, and K. Schwingenschuh, Model amplitude spectra of vertical electric and horizontal magnetic fields on Titan, *Planet. Space Sci., 51*, 852-862, 2003.

Nickolaenko, A.P., L.M. Rabinowicz, A.V. Shvets, and A.Yu. Schekotov, Polarization characteristics of low frequency resonances, Izvestija VUZOV, *Radiofizika, 47*, No.4, 267–291, 2004 (in Russian).

Nickolaenko, A.P., M. Hayakawa, M. Sekiguchi, Y. Ando, and K. Ohta, Model modifications in Schumann resonance intensity caused by a localized ionosphere disturbance over the earthquake epicenter, *Ann. Geophysicae, 24,* 567-575, 2006a.

Nickolaenko, A.P. M. Hayakawa, and M.Sekiguchi, Variations in global thunderstorm activity deduced from the OTD records, *Geophys. Res. Lett., 33*, L06823, doi:10.1029/2005 GL 024884, 2006b.

Nickolaenko, A. P. & M. Hayakawa, Recent studies of Schumann resonance and ELF transients, *J. Atmos. Electr., 27*, 19-39, 2007.

Nickolaenko, A. P., M. Hayakawa, T. Ogawa, and M. Komatsu, Q-bursts: A comparison of experimental and computed ELF waveforms, *Radio Sci., 43,* RS4014, doi:10.1029/2008RS003838, 2008.

Nickolaenko, A. P. & M. Hayakawa, Model disturbance of Schumann resonance by the SGR 1806-20 γ-ray flare on December 27, 2005, *J. Atmos. Electr., 30,* 1-11, 2010.

Ogawa, T., Y. Tanaka, T. Miura, and M. Yasuhara, Observations of natural ELF electromagnetic noises by using the ball antennas, *J. Geomagn. Geoelectr., 18, 443*-454, 1966.

Ogawa, T, Y. Tanaka, A.C. Fraser-Smith, and R. Gendrin, Worldwide simultaneity of a Q-burst in the Schumann resonance frequency range, *J. Geomagn. Geoelectr., 19*, 377-384, 1967.

Ogawa, T. & M. Komatsu, Analysis of Q-burst waveforms, *Radio Sci., 42*, RSs18, doi:10.1029/2006RS003493, 2007.

Ohta, K., N. Watanabe, and M. Hayakawa, Survey of anomalous Schumann resonance phenomena observed in Japan, in possible association with earthquakes in Taiwan, *Phys. Chem. Earth, 31*, 397-402, 2006.

Pasko, V. P. (Ed.), Recent advances in studies of Schumann resonances on earth and other planets of the solar system, Special section, *Radio Sci.,42*, RS2501-RS2520, 2007.

Pechony, O. & C. Price, Schumann resonance parameters calculated with a partially uniform knee model on Earth, Venus, Mars, and Titan, *Radio Sci., 39(5)*, RS5007, doi:10.1029/2004RS003056, 2004.

Polk, C. & F. Fitchen, Schumann resonances of the earth-ionosphere cavity-extremely low frequency reception at Kingston, R. I., *J. Res. NBS, 66D*, 313-318, 1962.

Price, C. & V. Mushtak, The impact of the August 27, 1998, γ-ray burst on the Schumann resonances, *J. Atmos. Solar-terr. Phys., 63*, 1043-1047, 2001.

Price, C., E. Greenberg, Y. Yair, G. Sátori, J. Bór, H. Fukunishi, M. Sato, M. Israelivitch, P. Moalem, A. Devir, Z. Levin, J. H. Joseph, I. Mayo, B. Ziv, and A. Sternlieb, Ground-based detection of TLE-producing intense lightning during the MEIDEX mission on board the space shuttle Columbia, *Geophys. Res. Lett., 31*, doi:10.1029/2004GL020711, 2004.

Price, C., O. Pechnoy, and E. Greenberg, Schumann resonances in lightning research, *J. Lightning Res., 1*, 1-15, 2007.

Rycroft, M. J., Low frequency disturbances of natural origin of the electric and magnetic fields of the earth, *Ph. D. thesis*, University of Cambridge, 1963.

Sao, K., M. Yamashita, S. Tanahashi, H. Jindoh, and K. Ohta, Experimental investigations of Schumann resonance frequencies, *J. Atmos. Terr. Phys., 35*, 247-253, 1973.

Sato, M. and H. Fukunishi, Global sprite occurrence locations and rates derived from triangulation of transient Schumann resonance events, *Geophys. Res. Lett., 30* (16):doi:10.1029/2003GL017291, 2003.

Sátori, G., On the dynamics of the north – south seasonal migration of global lightning, Proc. of 12th ICAE, Versailles, France, *Global Lightning and Climate*, pp. 1-4, 2003.

Sátori, G., E.R. Williams, and D.J. Boccippio, *On the dynamics of the North-South seasonal, migration of global lightning, AE32A-0166*, AGU Fall Meeting, San Francisco, December 8 – 12 , 2003.

Sekiguchi, M., M. Hayakawa, Y. Hobara, A. Nickolaenko, and E. Williams, Links of Schumann resonance intensity with average global land temperature, *Radiophysics and Electronics, 9(2),* 383 – 391, 2004 (in Russian).

Sekiguchi, M., M. Hayakawa, A.P. Nickolaenko, and Y. Hobara, Evidence on a link between the intensity of Schumann resonance and global surface temperature, *Ann. Geophysicae, 24*, 1809-1817, 2006.

Sentman, D.D., Magnetic elliptical polarization of Schumann resonances, *Radio Sci., 22,* 595 –606, 1987.

Shvets, A. V., A technique for reconstruction of global lightning distance profile from background Schumann resonance signal, *J. Atmos. Solar-terr. Phys., 63*, 1061–1074, 2001.

Shvets, A.V., A.P. Nickolaenko, G.G. Belyaev, and A.Yu. Schekotov, Analysis of variations of Schumann resonance parameters associated with Solar proton events, *Radiophysics and Electronics, 10*, No.1, 85 – 97, 2005 (in Russian).

Shvets, A.V., M. Hayakawa, M. Sekiguchi, and Y. Ando, Reconstruction of the global lightning distribution from ELF electromagnetic background signals, *J. Atmos. Solar-terr. Phys.,71*, 1405-1412, 2009.

Tanaka, Y. T., T. Terasawa, M. Yoshida, T. Horie, and M. Hayakawa, Ionospheric disturbances caused by SGR 1900+14 giant gamma ray flare in 1998: Constraints on the energy spectrum of the flare, *J. Geophys. Res., 113,* A07307, doi:10.1029/ 2008JA013119, 2008.

Terasawa,T., Y. Tanaka, Y. Takei, N. Kawai, A. Yoshida, K. Nomoto, I. Yoshikawa, Y. Saito, Y.Kasaba, T. Takashima, T. Mukai, H. Noda, T. Murakami, K. Watanabe, Y. Muraki, T. Yokoyama, and M. Hoshino, Repeated injection of energy in the first 600 ms of the gaint flare of SGR 1806-20, *Nature, 343*, 1110-1111, 2005

Williams, E.R., The Schumann resonance: a global tropical thermometer, *Science, 256*, 1184-1187, 1992.

Williams, E.R., V. Mushtak, and A.P. Nickolaenko, Distinguishing ionospheric models using Schumann resonance spectra, *J. Geophys. Res., 111*, D16107, doi:10.1029/ 2005JD006944, 2006.

Yamashita, K., T. Otsuyama, Y. Hobara, M. Sekiguchi, Y. Matsumoto, M. Hayakawa, and V. Korepanov, Global distribution and characteristics of intense lightning discharges as deduced from ELF transients observed at Moshiri(Japan), *J. Atmos. Electr., 29*, No.2, 71-80, 2009.

In: Lightning: Properties, Formation and Types
Editor: Matthew D. Wood
ISBN: 978-1-61122-198-5

Chapter 4

EDUCATIONAL COMPUTER SIMULATIONS OF LIGHTNING AND ASSOCIATED THUNDERS

***Francesco Blanco*[1,2] *, Paola La Rocca*[1,3]**
***Catia Petta*[1] *and Francesco Riggi*[1]**
[1]Department of Physics and Astronomy, University of Catania and INFN, Via S.Sofia 64, I95123 Catania, Italy
[2]University of Houston, Houston, TX, USA
[3]Museo Storico della Fisica e Centro Studi e Ricerche "Enrico Fermi", Roma, Italy

1. INTRODUCTION

1.1. Historical Background

Lightning and thunder have been observed since a long time in the history of man. Speculations on the origin and cause of thunder date to the ancient Greece, with the opinion of Aristotle, relating them to sudden ejections of vapours from clouds. Lucretius in his "*De rerum natura*" subsequently attributed the thunder to the collisions between clouds. During the next centuries, the explanations concerning the cause of thunder made no reference to the existence of lightning in the sky. At the beginning of 1700, W. Wall and I. Newton commented that the origin of lightning could be similar to that of crackling of sparks produced by rubbed amber rods. Benjamin Franklin, in 1749, was probably the first one to suggest the electrical nature of thunderstorms, and to carry out experimental investigations aiming at demonstrating that thunderclouds were charged, and that a large potential difference existed between clouds and the Earth.

Only at the beginning of the 20th century a general agreement arose on the nature of thunder, as being due to a shockwave in air, produced by the sudden expansion along the lightning channel. C.T.R.Wilson stressed the role of lightning in the charge transfer between the upper atmosphere and the Earth. The important role of the particular geometry of the lightning channel, with its tortuosity, and the time duration of the sound, was first pointed out by the physicist R.Hooke. Even though different theories concerning the specific mechanism of the generation process existed, a general agreement of the cause producing a thunder was reached by mid-nineteenth century. M.Hirn proposed in 1888 a simple picture of the thunder mechanism, as being due to the heating of the air and its sudden expansion due to the passage of electrical current.

A complete review of thunder research from the old era to 1960 has been compiled by Remillard [1]. Since that time, there has been a growing scientific interest in the study of thunders, partly due to the availability of new technologies which helped to elucidate the detailed aspects of the lighting and the associated sounds.

1.2. The Lightning Mechanism

A lightning flash usually has its start at the base of a cloud in a region of large negative charge. The altitude at which this happens may typically reach 5 km, where the temperature is approximately -10 °C. In that environment, water droplets freeze, and this process may be connected with the generation of electric charge.

The lightning flash consists of several strokes. Even though the potential difference between the ground and the clouds can be as high as hundreds of million volts, this is not enough to drive a single arc across a distance of several km in air. The main discharge can only happen when the channel has been traced by a low current discharge (usually called the stepped leader), associated to electrons accelerated by the electric field, producing more electrons in their collisions with air molecules, and resulting in a conductive path of partially ionized air. The stepped leader may form numerous branches and it takes of the order of 20 ms to create the channel. Once a conductive path has been formed, a large current (20000 A or more) may flow from the ground to the cloud, which constitutes the return stroke. Such channel is the most luminous part of the flash, and its diameter is very small (few mm). This core may reach a temperature which can exceed 3×10^4 K. The spectral composition of the lightning flashes has been extensively studied by spectroscopic investigations, and several spectral lines were found in the infrared, visible and ultraviolet regions [2].

A typical lightning flash may have three or more leaders, each followed by a return stroke. The time structure of the overall lightning geometry is very short on the time scale of the associated thunder, and what is relevant to determine the sound is the spatial distribution of the different segments constituting the channel. The shape of a typical lightning channel is tortuous, consisting of straight segments separated by sharp bends. The tortuosity of the lighting channel exhibits three different scales: the first one, with straight segments at least 100 m long, an intermediate scale with segments between 5 and 100 m, and a microtortuous scale, with segments smaller than 5 m. Such specific structure has a deep influence on the sound being produced.

1.3. The Sound of a Thunder

The increase of the pressure inside a small volume, up to 10-1000 atmospheres [2], is the cause of a shock wave. In a few meters from the source, the shockwave dissipates most of its energy into heat. However, a small fraction still remains, producing a normal acoustic wave which propagates in the air at the speed of sound. Since the energy originally associated to the shockwave is very large, even a small fraction of it – in the order of 1 % - is still able to produce sounds which are audible at km distances. The detailed shape of the shock wave is very complex, and it changes in a complicated way as it expands. However, at a short distance from the original source volume (the relaxation radius, in the order of a few meters) the structure of the pressure wave may be described in a relatively simple way, which is explained later in this Chapter.

Qualitative and quantitative observations of the sound of thunder have been long reported [2]. Qualitative descriptions generally identify two different phases of a thunder. The first one – a 'crash' – is very intense, especially close to the point where the lightning hits the ground; a second one – a long 'roll' or rumble – is generally associated to the particular way through which the sound is propagated and delayed due to the extended source and transmission channel. This explains why different observers, located far away each other, do not perceive the same sound pattern.

As far as the sound spectrum of a thunder is concerned, several detailed measurements have been carried out [2] from the beginning of 20th century to recent years, with a variety of experimental techniques. Since each thunder has its peculiar pattern, investigations generally agree only on the fact that its power spectrum is broad, in the range from a few Hz to a few hundred Hz, with a peak at a frequency in the order of 100 Hz.

The propagation of the individual acoustic waves from each portion of the lightning channel to the observer involves several aspects. They determine the shape of the pattern which may be heard in a fixed position, as well as the maximum distance for a thunder to be heard.

As an example, Figure 1 depicts a basic geometry for a lightning stroke which hits the ground at (x=0, y=0). A potential observer, located at (x = -1750 m, y=0) receives sounds from any portion of the lightning channel, with a specific delay determined by the distance between this portion and the observer. The intensity of each contributing sound is influenced by the particular orientation of the lightning segment and also on its distance to the observer, which may attenuate the sound.

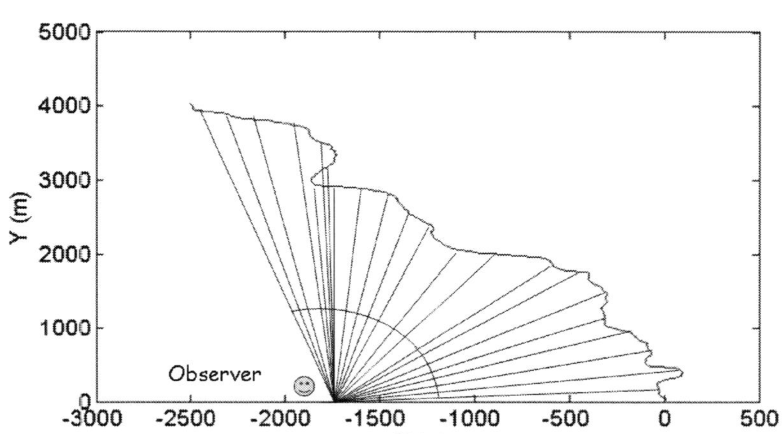

Figure 1. An example of lightning geometry, with the various segments contributing to the overall sound pattern heard from an observer.

The evaluation of the overall sound pattern perceived by the observer then includes the original power spectrum associated to each portion of the extended source, the delay introduced by the distance travelled by each wave, the detailed geometrical pattern of the lightning channel – with its ramifications and structure – and the attenuation of the acoustic wave, which in general depends on the frequency of each component. Finally, the temperature and density profile of the air in the atmosphere may also induce refraction processes of the waves, which are the analog counterpart of the optical mirages. All these factors are very complicated to take into account, and may require powerful simulations of the different aspects, which have been pioneered by Ribner and Roy [3,4] and revisited in recent years by Few [5] and Glassner [6].

From an experimental point of view, the different signatures of a thunder as detected at different observer positions, may be investigated by an array of microphones properly placed on flat ground [5]. It was verified that until two microphones are separated by a horizontal distance of 20 to 30 m, the thunder signatures are similar, with only minor differences. When the microphones are placed 100 m apart, most of the details are different, even though the underlying pattern is preserved. Finally, with microphones separated by more than a few km apart, thunder signatures are totally different. Moreover, even the height above ground may influence the signatures recorded by a microphone, such that a few meters displacement may result in audible differences. The complete reconstruction of a lightning should then require a large array of microphones sensitive to low frequency sounds.

1.4. Educational Aspect of the Present Work

The simulation of digital thunders, following a realistic model of the lightning geometry and the superposition of the elementary acoustic waves up to the production of the resulting sound pattern as heard from an observer, may have powerful educational aspects. When discussing in an introductory physics course the basic aspects of sound, such as its propagation and speed, a common example is often presented, i.e. the time elapsed from the visual observation of the lightning flash in the sky to the first sound being heard. This allows to estimate the approximate distance at which the electrical discharge occurred. However, this only gives an idea of the minimum path the sound had to travel to reach our ears, and cannot explain why the thunder lasts for several seconds, with a complicated and specific pattern.

To carry out a simulation of the lightning and thunder structure for educational purposes even by means of simplifying hypotheses and widely available computational tools, without making use of sophisticated models and dedicated software, we employed the well known MATLAB® package. This was used not only to model the geometry of the lightning channel and to generate and propagate in air the acoustic waves from such extended source, but also to plot the results and play the corresponding sound pattern at the end of the simulation procedure. Although very simple, the model allows to modify several parameters, to understand how they alter the resulting sound pattern. The description of this activity is intended as a suggestion to physics students and teachers at the university level, to provide them with an educational tool and advanced examples when introducing the concepts of sound emission and propagation. While the basic model may be discussed even in introductory university physics courses, additional improvements of such approach – as outlined in the concluding remarks – may constitute a subject for more advanced activities in higher education.

2. The Simulation Model

The first step in the simulation is to build a realistic geometry of the lightning channel by some mathematical algorithm. Looking at real pictures of a lightning, one can realize that even a single branch lightning has a complex structure, made by a tortuous channel with several segments. Visual observation of real pictures taken during thunderstorms has allowed in the past to extract basic information on the geometry of such lightning channels. As discussed in the Introduction, it was seen that they exhibit a large scale (> 100 m), an intermediate scale (5-100 m) and a fine scale (~3 m) tortuosity, the latter being more important to produce realistic thunder signatures. To carry out such simulations, it is possible to follow the approach described in Refs.[3,4], modelling the lightning as a two-dimensional (or three-dimensional) random walk originating from a point at ground (assumed as the origin of the reference system), and described in terms of the (x,y) coordinates which mark the ends of each segment. Although the true structure of a real lightning is a 3D geometry, even a 2D random walk has enough details to make the results interesting, with a non negligible simplification of the mathematics required. Moreover, understanding the connection between the detailed geometrical structure of the lightning and the main features of the thunder pattern is easier with a 2D geometry, and more effective for educational purposes.

To reproduce the complex structure of the overall lightning channel, the segments constituting the channel must be assumed as having different lengths. In our simplified model, their distribution was assumed to follow a simple exponential function

$$P(l) \sim \exp(-l/l_{ave}) \tag{1}$$

or a distribution of the form

$$P(l) \sim l \exp(-l/l_{ave}) \tag{1'}$$

with an average value $l_{ave} = 3$ m.

The overall number of segments constituting the lightning stroke determines the global length of the channel. Several trials were made with a number of segments in the order of 1000-2000, giving lightning channels which extend from ground to altitudes up to 4000 m. To model the ramification of the lightning tree, several tests were also made concerning the choice of the opening angle $\Delta\theta$ between two adjacent segments. In agreement with what discussed in Ref.[3], we found that a simple Gaussian distribution

$$P(\Delta\theta) \sim \exp[\,(-\Delta\theta/\Delta\theta_m)^2\,] \tag{2}$$

with $\Delta\theta_m$ in the order of 10°-30°, is not reasonable, since quite often it produces a lightning geometry which exhibits an inversion of the original direction, with lightings which head towards the ground again.

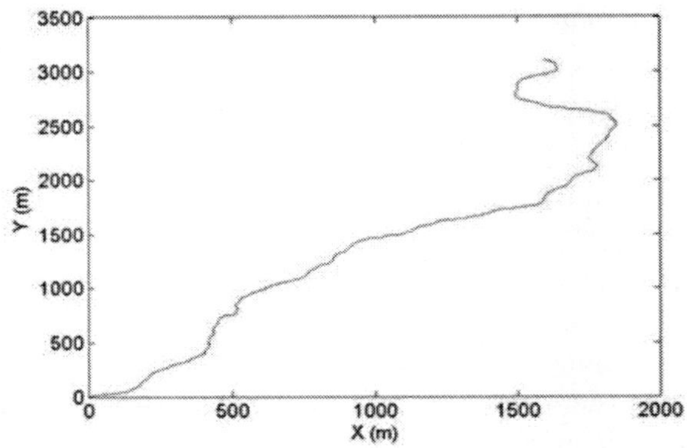

Figure 2. A lightning geometry, simulated by the algorithm described in the text.

Table I.

For sin θ < φ :		
	P(t) =0	τ < -φ –sin θ
	(-B/sin θ) [(τ+sin θ)^2 – φ^2]	-φ - sin θ < τ < -φ + sin θ
	- 4 B τ	-φ + sin θ < τ < φ - sin θ
	(B/sin θ) [(τ - sin θ)^2 – φ^2]	φ - sin θ < τ < φ + sin θ
	0	φ + sin θ < τ
For sin θ > φ :		
	P(t) = 0	τ < -φ –sin θ
	-(B/sin θ) [(τ + sin θ)^2 – φ^2]	-φ - sin θ < τ < φ - sin θ
	0	φ - sin θ < τ < -φ + sin θ
	(B/sin θ) [(τ - sin θ)^2 – φ^2]	-φ + sin θ < τ < φ + sin θ
	0	φ + sin θ < τ

where $\tau = (v\,t - r)/l$, $\varphi = v\,T/l$, $B = l^2/(2\,r\,v\,T)$

v = sound speed

l = length of each segment

r= distance from the observer

T = duration of individual N-waves

θ = angle between the normal to the segment and the line from the center of the segment to the observer

Using a similar approach to that outlined in [3], some memory of the initial lightning direction was introduced in the distribution, biasing the angle extracted from the Gaussian spreading with the average direction of the last few segments. After some trials, example of lightnings which are not too unrealistic could be obtained, as depicted in Figure 2.

The geometrical structure of the lightning channel described by the above procedure is at the origin of the elementary waves emanating from each point along the channel. Instead of generating elementary acoustic pulses from each point, a commonly adopted approach is to analytically integrate the contributions within each segment along the channel, taking into account the orientation of that particular segment with respect to the observer. Such approach was originally designed by Wright and Medendorp [7] and successfully employed in Refs.[3-

4,6]. Considering a finite straight linear segment, and assuming each point along the segment as the source of an "N-wave" (a sudden compression followed by a decompression), the resulting wave obtained by analytical integration of N-waves along the segment may be expressed by the relations shown in Table I [3].

For a given observer position, the time history of the pressure signal may then be obtained by a sum of the various contributions of all individual segments, with a proper choice of the relevant parameters (geometrical position of the observer, duration T of the elementary waves, average length of the lightning segments). Depending on the angle orientation θ between the normal to each segment and the line connecting the centre of the segment to the observer, Figure 3 shows how each contribution may assume the form of an exact N-wave (for θ = 0), a round N-wave, or two distinct parabolas.

The addition of the individual waves emanating from each segment gives the resulting signal. Taking into account the distance d travelled from each segment to the observer, an attenuation factor exp(-d/D) was also introduced for each wave, to take into account the sound absorption in air.

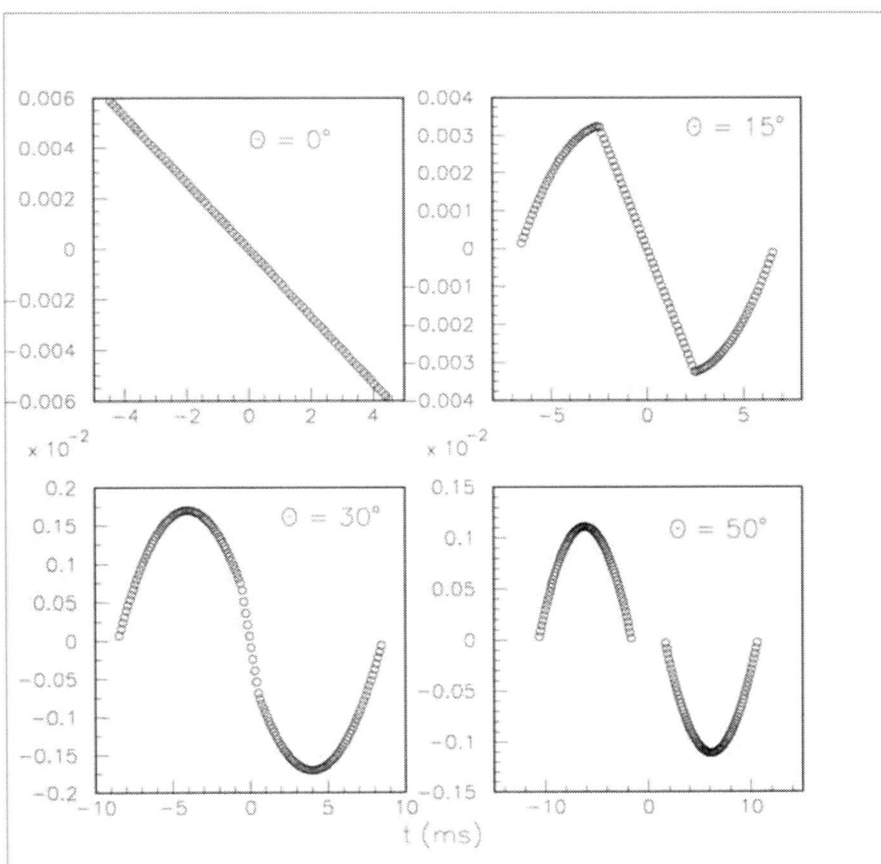

Figure 3. Evolution of the shape of elementary waves emitted by the various segments of the lightning stroke, for different orientations with respect to the observer.

3. COMPUTER AND PROGRAMMING TOOLS

Digital sound patterns may in principle be generated through any programming language (Fortran, C++,..). Early approaches to the problem of reproducing a digital thunder [3] made use of the state-of-art computer technology available at that time (HP2100 or IBM370). The pressure-time history was sampled at fixed time intervals and the result stored on magnetic tape. After filtering it, the signal could be played through an amplifier-loudspeaker system, or transferred to a tape recorder for later use. Since the computer technology has considerably evolved over the last two decades, a variety of programming tools have become now available, which allow to implement all the necessary operations, from formulae evaluation to sampling the sound pattern, to graphical visualization of the lighting strokes and the sound pattern, to finally playing the obtained sound through the control of an audio card. One of these packages is MATLAB® [8], which is a high level language and interactive environment, allowing to develop algorithms, data analysis and visualization. It is used in a wide range of applications including numerical computation, signal and image processing. For such educational study we implemented all the necessary operations through MATLAB macros. The main macro was used to implement the formulae reported in Table I and to sample the pressure-time history at fixed time intervals, storing the result on appropriate files. Within a loop on the number of stroke segments, the orientation and length of each segment is randomly extracted according to Eqs.(1-2), in order to retain some memory of the initial direction by an average over the last four segments. The contribution to the amplitude is evaluated for each segment following the formulae reported in Table I, with an exponential attenuation factor depending on the travelled distance. The geometry of the lightning channel, the resulting amplitude-time pattern and the corresponding audio file are then saved at the end of the process. Different choices of the parameters may easily be introduced into the model, to modify the relevant parameters (default values reported in parenthesis). These include the position of the observer along the x-direction, the starting orientation of the lightning channel near the ground (usually close to the vertical), the number of segments constituting the tree structure of the lightning (around n=2000), the average length of the segments (of the order of l_{ave} = 3 m), the RMS of the angle spreading between consecutive segments (by default assumed as $\Delta\theta_m = 30°$), the time duration of the individual pulses (T = 5 ms), the speed of sound (v = 340 m/s), and the attenuation length (D = 500 m). The CPU time to simulate a digital thunder with such default parameters is of the order of one minute on a recent processor. Different random number sequences may be used to produce different geometries of the lightning channel, by a proper choice of the random seed. Additional small macros were then employed to plot the 2D structure of the lightning channel and the resulting sound pattern, as well as to convert the audio files into any specific audio format and to play them.

4. RESULTS

Each lighting geometry produces its peculiar series of thunder sound patterns. The geometry of the channel constitutes the first degree of freedom in the evaluation of the final result, since a large variety of shapes may actually be produced even from simple algorithms, such as those discussed here.

Figures 4 and 5 show two typical geometrical patterns which can be obtained with realistic choices of the parameters discussed in Sect.2.

One of the main results coming out from computer simulations of thunders originating from a lightning geometry is the large variety of sound patterns which can be produced even with a small change in the position observer. Using the same lightning geometry (shown in Figure 4), we produced several examples of sound patterns (Figs. 6-12) for different positions of the observer with respect to the place where the lighting hits the ground (a very dangerous place indeed), from Xobs= -10 m to Xobs = -2500 m.

The 2D lightning stroke depicted in Figure 4 extends from ground (y=0) to about y=4000 m, spanning an area of more than 10 km^2 over such plane. It has enough details to produce a complicated and position-dependent sound pattern. Figure 6 shows the pattern which is expected very close to the point where the lightning hits the ground (Xobs=-10 m, y=0). A very intense pulse is seen at time t=0, followed by a smaller intensity sound which extends for the first 0.1-0.3 s; some additional bumps are seen at t~1 s and t~1.4 s, roughly corresponding to distances of 340 m and 480 m from the observer. The intensity of the second bump is about 10% of the main peak.

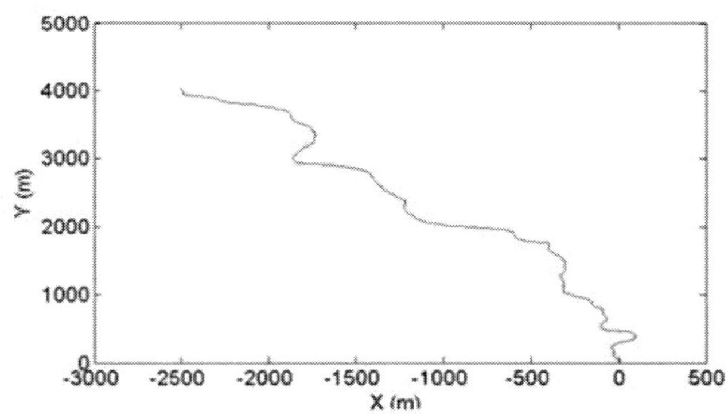

Figure 4. A geometrical pattern of the lightning stroke.

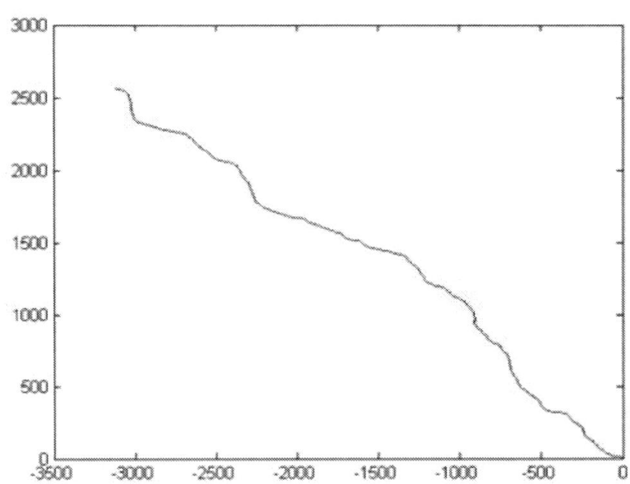

Figure 5. Another example of lightning geometrical pattern.

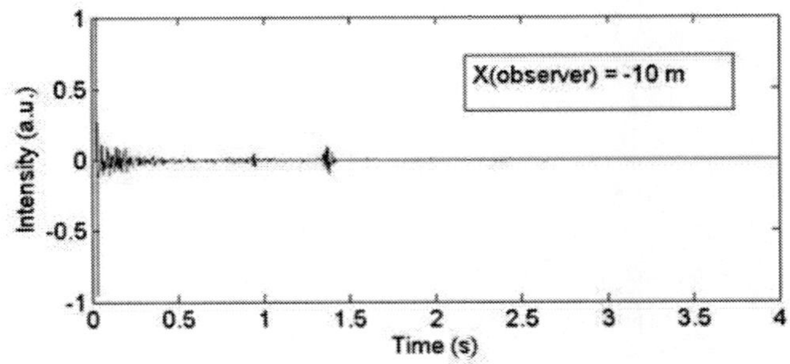

Figure 6. Sound pattern as heard from an observer located at x= - 10 m with respect to the place where the lightning stroke hits the ground.

Moving the position of the observer even by a small quantity (from Xobs=-10 m to Xobs=-50 m), Figure 7, gives a different pattern, since the amplitude ratios between the main peak and the satellite peaks at t=1 s and t=1.4 s are largely modified. The main peak is now shifted at about 0.2 s and the peak at 1.4 s is now about 20% of the main peak. These features are even more apparent (Figure 8) by a further shift of the observer position (Xobs=-100 m), where the main peak occurs at 0.3 s and the secondary peak at 1.4 s is now 60% of the main peak amplitude. It must be noted that the sound starts now at 0.2-0.3 s after the light hit, since the time delay starts to be non negligible.

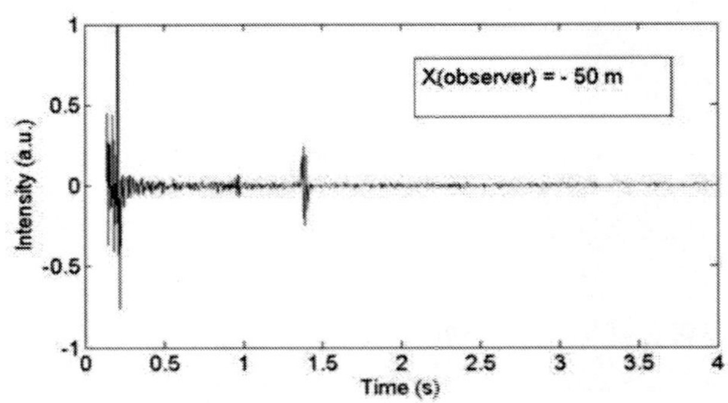

Figure 7. As for Figure 6, for an observer located at x= - 50 m.

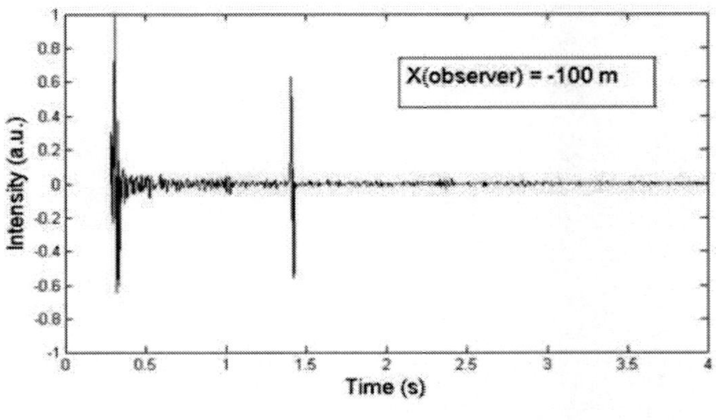

Figure 8. As for Figure 6, for an observer located at x= - 100 m.

Going even further, to Xobs=-250 m (Figure 9), the sound starts at about 0.8 s with a second peak at 1.4 s (about 40-50% of the main one). In such conditions, a larger background sound is however present before and after the second peak, which extends to longer times (5-6 s, corresponding to portions of the lightning stroke distant more than 1500 m). This part of the sound pattern is mainly due to the portion of the lightning stroke located between 2000 and 3000 m altitude. Placing the observer at Xobs=-500 m (Figure 10), the relative importance of these portions increases, since the sound coming from the lightning portions near the ground starts to be attenuated, so that the amplitudes of the sound patterns at 2-3 s and 5-6 s are comparable. The zone between 3 and 5 s is relatively "quiet", since no many lightning segments are located at distances 1000-1600 m.

An additional shift of the observer position to Xobs=-1000 m (Figure 11) produces a sound pattern starting at about 3 s after the light hit and whose main features last for about 3 s, with a longer rumble after that. Finally, at Xobs=-2500 m (Figure 12), most of the sound pattern is concentrated between 7 and 12 s, with several comparable peaks which are the effect of the specific structure of the lightning channel.

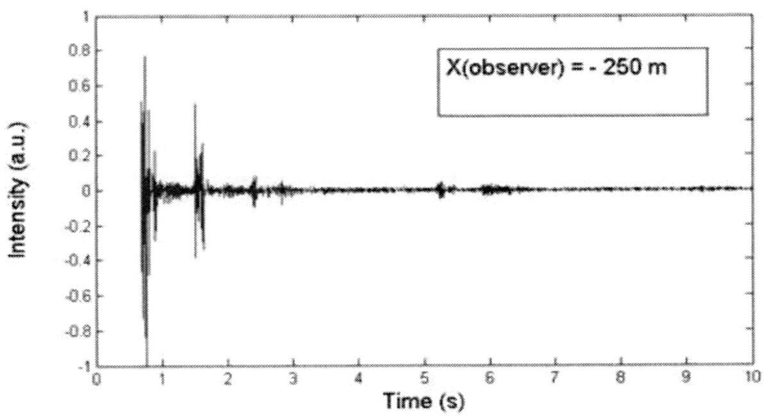

Figure 9. As for Figure 6, for an observer located at x= - 250 m.

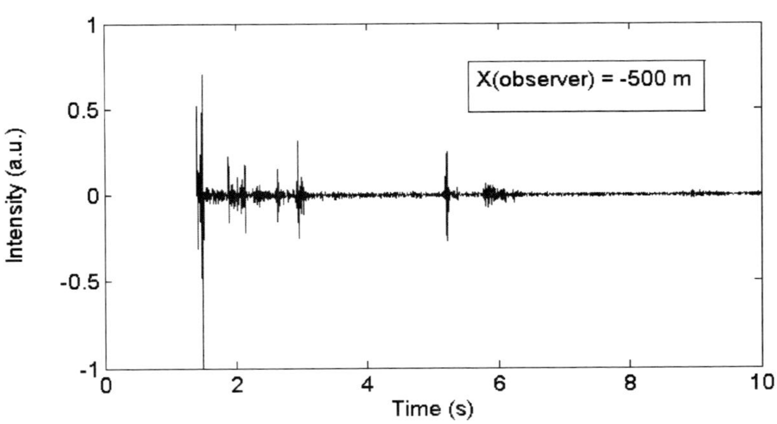

Figure 10. As for Figure 6, for an observer located at x= - 500 m.

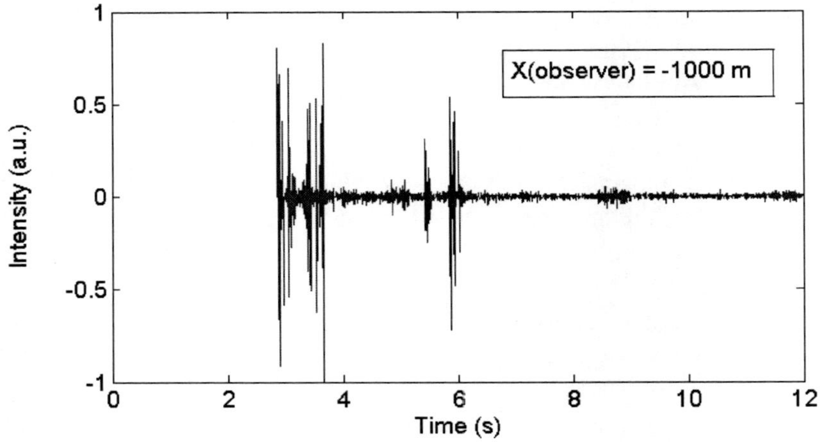

Figure 11. As for Figure 6, for an observer located at x= - 1000 m.

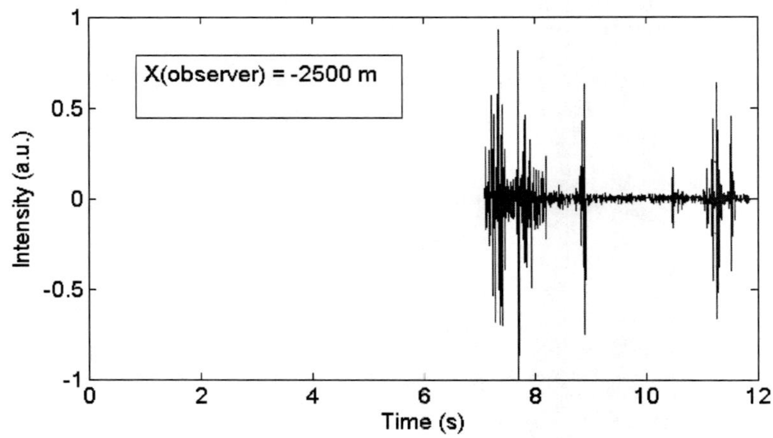

Figure 12. As for Figure 6, for an observer located at x= - 2500 m.

5. Conclusion

An implementation of a realistic yet simple model for the generation of lightning channels and associated sound patterns has been carried out. The results from the simulation have confirmed that choosing different topologies of the simulated lightning channel and/or different observer positions produces a variety of sound patterns, even by a simplified 2D approach.

The present model, which was employed in the past for more detailed investigations, was shown to be of educational help to understand, with a few basic hypotheses, how the complex structure of the sound pattern associated to a lightning originates from its geometry. A widely available software package was used as a stand-alone tool to model the random geometry of a lightning channel and produce the corresponding sound pattern. Additional improvements of the method are possible along several directions. More realistic lightning geometries may be modelled by a 3D approach, and additional ramifications along the main branch may be added. A lightning path inside clouds, with its proper ramifications, could add realism to the description of the lightning structure. As far as the attenuation of the sound waves in air is concerned, a detailed dependence of the attenuation factor on the frequency components may be introduced, as well as filtering effects (low pass filters) which modify the original sound spectrum. The propagation of the sound waves could also be modified introducing refraction (due to the atmosphere density and temperature profiles) and reflection effects (due to the ground or large obstacles). All these improvements tend to complicate the basic model in order to arrive at a description of the final sound pattern as realistic as possible, and could be the subject for additional educational investigations on this topic. In this respect, it may be relevant for students to understand how even a complex natural phenomenon may be modelled by mathematical and physical descriptions which make use of different levels of approximation. A previous paper on this subject has been already published by the authors [9].

REFERENCES

[1] Remillard, W. J. (1961). *Weather*, *16*, 245.
[2] Graneau, P. (1989). *J.Physics*, *D22*, 1083.
[3] Roy, D. (1981). A Monte Carlo Model of Tortuous Lightning and the Generation of Thunder, PhD dissertation, UTIAS Techn. Report no.243, *Institute for Aerospace Studies*, Univ.of Toronto.
[4] Ribner, H. S. & Roy, D. (1972). *Journal of Acoustical Society of America, 72*, 1911.
[5] Few, A. A. (1975). July, *Scientific American, 80*.
[6] Glassner, A, (2000). *IEEE Computer Graphics and Applications, Part I* (March 2000) and Part II (May).
[7] Wright, W. M. & Medendorp, N. W. (1968). *Journal of Acoustical Society of America, 43*, 966.
[8] http://www.mathworks.com/.
[9] Blanco, F., La Rocca, P., Petta, C. & Riggi, F. (2009). *Eur. J. Physics*, *30*, 139.

In: Lightning: Properties, Formation and Types
Editor: Matthew D. Wood
ISBN: 978-1-61122-198-5

Chapter 5

Lightning in Protoplanetary Disks

Takayuki Muranushi *
The Hakubi Center, Kyoto University, Japan;
Yukawa Institute for Theoretical Physics, Kyoto University,
Sakyo-ku, Kyoto, 606-8502, Japan

PACS 05.45-a, 52.35.Mw, 96.50.Fm.

Keywords: Lightning, Protoplanetary Disks, methods:numerical — planetary systems: formation — planetary systems:protoplanetary discs — meteors, meteoroids — plasmas — turbulence.

AMS Subject Classification: 53D, 37C, 65P.

1. Introduction

Why are some astrophysicists concerned with lightning? Because the current understanding of the planet formation theory has several problems, such as inconsistency, and missing pieces of the puzzle. By the way, although the effect of electric charge of the dust aggregates has been strangely neglected by most astrophysicists until recently, it has actually been found to be a crucial factor. If we consider the global motion of such charged dust aggregates, the generated electrostatic field is sometimes large enough to cause lightning.

To share our current understandings and problems with you, let us first introduce you to the theory of planet formation, which takes place in protoplanetary disks.

1.1. Protoplanetary Disks

What are the protoplanetary disks and how do they form? In our galaxy there are stars and planets and some other weird objects, and the vast space between them is filled with thin, thin gas. The gas is typically either atomic or ionized hydrogen, its number density smaller

*E-mail address: nushio@yukawa.kyoto-u.ac.jp, muranushi@gmail.com

than 1 cm^{-3}. Less than 1% in volume, but about half the mass of the interstellar gas, are clouds of molecular hydrogen H_2. From such molecular clouds protoplanetary disks form.

There also are various molecules and minerals in the form of tiny solid particles. Their typical radius is $0.1\ \mu\mathrm{m}$. Astronomers often call the small solid particle "dust." The term applies to any solid particle small, in the astronomical sense; thus we sometimes say "a km-sized dust." It is in fact a meteorite as large as a mountain. Although, we do not always imagine a piece of stone when we say "dust;" because there is no gravity in space, when two pieces of relatively small "dust" collide, they silently stick together and form a fluffy, very low-density structure that resembles, yes, house dust, or a piece of cotton. A typical dust aggregate with $1\ \mathrm{m}$ radius is lighter than the same volume of 1 atmosphere pressure air. In a protoplanetary disk, such dust, starting from $0.1\ \mu\mathrm{m}$ in radius, takes millions of years to stick together, and finally become thousands of kilometers in radii — those, we call planets, on one of which we dwell.

The study on the origin of the solar system started in the 1960s, independently by Japanese, U.S. and Russian group, respectively led by Hayashi, Cameron, and Safronov. Based on Hayashi et al.'s[18] work is today's "standard scenario of planet formation." c.f. Figure 1 Some parts of the scenario are known to be problematic today, and observational results are coming in that are beyond the imagination of the scenario. We introduce such problems in §1.2.. Nevertheless such updates are patchy and no astrophysicist has succeeded in combining those new results into a new, self-consistent scenario. Hence we introduce the standard scenario, whose significance as a theoretical backbone is still undamaged.

- **Stage 1.** Protostar and protoplanetary disk formation.

  The densest part ($\sim 10^6\ \mathrm{cm}^{-3}$) of a molecular cloud further cools and contracts by its self-gravity, and becomes a protostar, in the beginning as heavy as a Jupiter. The remainder of the gas ($1 \sim 100 M_\odot$; $M_\odot$ is a solar mass,) cannot fall immediately to the star because of its initial angular momentum. Thus it forms a rotating disk-shaped object around the protostar. This is a protoplanetary disk.

- **Stage 2.** Protostar growth.

  The disk generates density waves due to its self-gravity. The waves carry the angular momentum outside, and the disk material is possible to accrete onto the protostar. The star shines due to the released gravitational energy, and spews out stellar wind. The wind limits the accretion at some point and fixes the stellar mass. The disk mass decreases gradually, and when it is about $0.01 M_\odot$, planet formation starts.

- **Stage 3.** Dust sedimentation and formation of the protoplanets.

  As the disk becomes not as turbulent as before, dust particles can stick together and form aggregates. The grown dust sediments to the disk equatorial plane. When enough dust is gathered on the equatorial plane, the dust swarm collapses by its self-gravity to form km-sized protoplanets.

- **Stage 4.** Protoplanet growth through collision

  Protoplanets collide and merge together. The growth is oligarchic [21]: larger planetesimals can gather more material and grow faster. Thus in the end, dozens of big

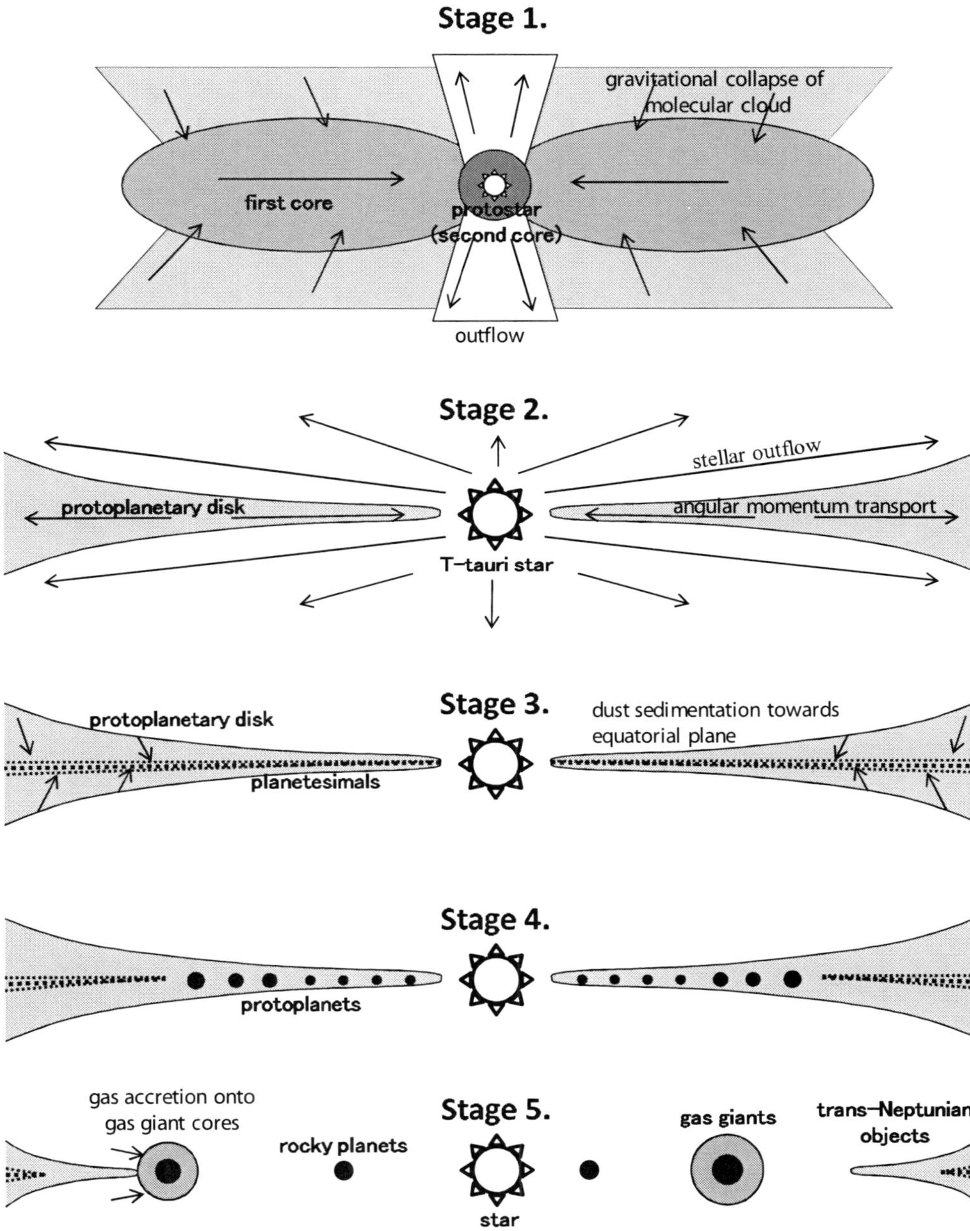

Figure 1. The "standard scenario" of planet formation

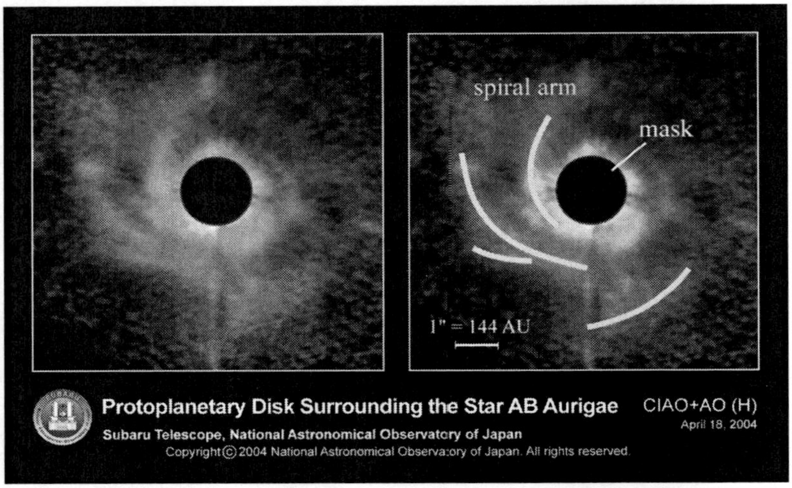

Figure 2. Picture of a protoplanetary disk, taken by Subaru telescope. Original images can be obtained from http://subarutelescope.org/Pressrelease/2004/04/18/index.html .

planetesimals are formed. Their mass was $0.1M_{\oplus}$ around the earth orbit and $2M_{\oplus}$ around the Jupiter orbit ($M_{\oplus}$: an Earth mass.)

- **Stage 5.** Gas giant formation and the end of the disk

  Heavy planets formed among the Jupiter and Saturn orbit accrete protoplanetary disk gas until they deplete the gas around their orbit and open gaps in the disk. Thus they become gas giant planets. Lighter planets, which are formed around the inner orbits, are not large enough to cause gas capture. After a few final giant impacts, they become rocky planets like Mercury, Venus, Earth, Mars. Eventually the disk gas disappears and a planetary system is formed. At outer orbits where material is scarce, the rocky cores grow slowly and do not accrete enough gas, leading to not-fully-mature gas giants like Uranus and Neptune. At even outer orbits, unused planetesimals remain for billions of years, as the Kuiper belt and Oort cloud.

1.2. Problems of Planet Formation Theory

This standard scenario is an elaborate work, and used as a foundation of today's planet formation theory. However, later on, several problems are found on the standard scenario.

To understand these mechanisms, it helps to note that dust(solid) and gas component has different orbital velocity. The orbital velocity of a point mass around a star is determined by the balance between the gravitational force from the central star and the centrifugal force from the orbital rotation. The velocity is called Kepler velocity. The gas component has the sub-Keplarian orbital velocity because the pressure gradient (the closer to the central star, the higher is the pressure) assists the centrifugal force. The solid component is not affected by this pressure gradient and tends to orbit in Kepler velocity. The heaviest bodies ($r > 1$ km) do so and due to their inertia their orbit is unaffected by the gas. The lighter planetesimals ($1 \text{ cm} < r < 1 \text{ m}$) loses their angular momentum to the slowly-rotating gas via gas friction, and tend to fall into the star. The even lighter, or lower density dust, is coupled to the gas motion via friction and do not fall.

- Prohibition of protoplanet formation by Kelvin-Helmholtz instability.

  As the dust sediment at the disk equatorial plane (Stage 3), the relative velocity between the equatorial dust layer and high-altitude gas layer become greater. The differential velocity causes a Kelvin-Helmholtz instability which will disturb the dust-sedimented layer. Thus dust layer cannot reach the dust density enough to cause self-gravity-driven planetesimal formation.

- Loss of cm-sized dust by migration.

  There is an alternative scenario for planetesimal growth (Stage 3): planetesimals can gradually grow by mutual sticking. In this case, when planetesimal size reaches $\sim 1 \text{ cm}$, they begin to migrate towards the central star faster than they grow and cannot grow further.

- Loss of earth-sized protoplanet by type-I migration.

  Even if the planetesimals manage to grow larger than 1 km in size, when they reach the earth radius (Stage 4), they will raise spiral-arm structures in the protoplanetary disk. The arm delivers the angular momentum from the planet to the disk; this cause the young planets to fall into the star before the protoplanetary disk disappears.

- Suggestion from observations.

  One of the latest progress in astronomy is that we are finally possible to observe the direct images of extra-solar planets, that is, planets that orbit stars other than the Sun [20, 27, 25, 47]. However, those observed gas-giant planets have orbital radii of hundreds of astronomical units (AU), while the standard model expect such planets to be in orbit $2 \sim 20$ AU. Also, direct image of protoplanetary disks that extend to hundreds of AU are available now (e.g. Figure 2.) These images, enabled by progresses in observation techniques, bring in indispensable information on planet formation. Since some of them are totally unexpected and intriguing, astrophysicists are now forced to reconstruct, or, enrich their theory on planet formation.

One of the big assumption of the standard scenario is that planets are formed near where they are today; in this sense it is a stational model. Observations mentioned above cast questions to this assumption. Planets may have formed in wider area, compared to today's solar system, and dynamism in protoplanetary disk brought them to where they are today. Dynamism of the disk is exiting topic of study today. Dynamism, to which electromagnetic processes, along with hydrodynamic and gravitational processes, contribute.

1.3. Electromagnetic Processes in Protoplanetary Disks

Figure 3 depicts the several elemental processes related to electricity or magnetism proposed to be working in protoplanetary disk.

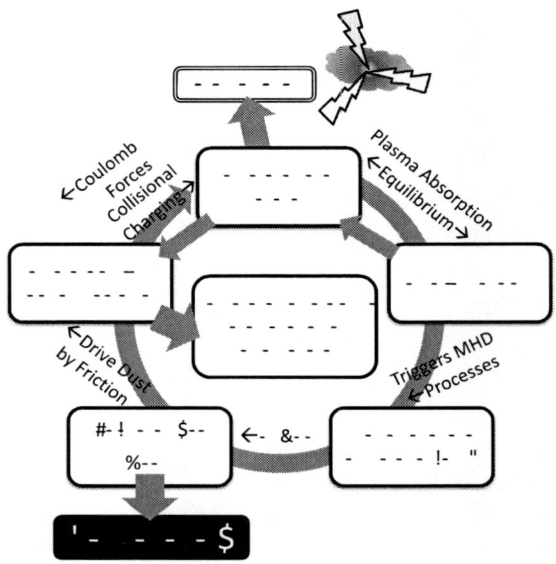

Figure 3. The cycle of electric and magnetic processes that affect each other. They are responsible for the fate of both the planets and the protoplanetary disk. Lightning (or, absence of it) is one of observational channels unique to electromagnetic processes.

One of the well-known roles of the dust in protoplanetary disk is to absorb charged plasma particles. The ionized degree of the disk is determined as equilibrium between the dust absorption and ionization processes. The main ionization source is cosmic rays, while radioactive elements have second-important contributions [49]. Gas ionization is the factor that determines whether the gas behaves like a fluid or magnetized fluid. This is most important because the magnetorotational instability (MRI), a magnetohydrodynamic process, is the most promising explanation for disk turbulence [1, 2]. Predicted ionization ratio in protoplanetary disk is dangerously low so that it can be partially or mostly MRI inactive [13]. When active, MRI powers the disk turbulence and wind. Turbulence causes relative motion among the dust aggregates because each aggregate respond differently according to its inertial mass or aerodynamic shape. Such relative motion cause dust aggregates to collide each other. Collisions can either grow or destroy the dust. And collision also contribute to dust charge state, through triboelectric charging or collisional charge separation.

The 9 o'clock to 12 o'clock arrows in Figure 3, that the dust charge can actually affect dust motion or even global dynamics of the disk, was not considered seriously until recently. The fact that dust-dust collisional charging is non-negligible closes the circle, and suggests that a comprehensive treatments are necessary for successful planet formation theory. It also suggests a unique means of observation: lightning. Although these electromagnetic processes surely affect other well-established observational signals such as dust mass-distributions or disk spectrum-energy diagrams, such signals are confused from non-electromagnetic processes such as disk turbulence, radiative transport etc. If disk lightning does exist and is observable, we will have a separate channel that carries the information on

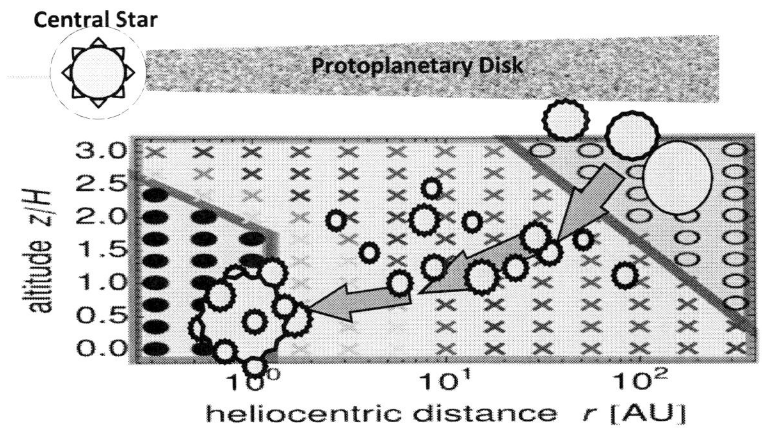

Figure 4. Okuzumi [33]'s estimate on possibilities of planetesimal formations. $\times$: dust aggregate growth is prohibited by electrostatic repulsion at the very early stage; $\bigcirc$: dust growth is permitted; $\bullet$: spontaneous bimodal dust growth. The regions ($1 \sim 100$ AU) assumed in the standard scenario as the main sites of planet formation are almost entirely marked by $\times$; i.e. planet formation is impossible, unless "seeds" are provided from elsewhere.

undoubtedly electromagnetic processes.

The electromagnetic nature of the processes is fundamental to update the "Standard Scenario," and to answer the two most basic questions of planet formation theory: How did planets form? How and when did protoplanetary disk disappeared? We know that there was a processes that put an end to protoplanetary disk, because it is not seen at our solar system today. It could have dumped into the sun, or lifted above the disk by disk-wind and then evaporated. We do not know how efficient the two processes were, but we know both are driven by magnetorotational instability. Also recently we found that electrostatic charge of the dust aggregates will alter the initial stages of planet formation scenario. Let us first look at this point.

2. Dust Charge due to Plasma Absorption

The ionization degree of a protoplanetary disk is in equilibrium: The main ionization source of the plasma is cosmic ray impacts, creating free electrons and positively charged molecular ions from disk gas medium. Then what is the main recombination source? The gas phase recombination, where a free electron is captured by an ion and result in a neutral molecule, is inefficient due to low density of the medium. The majority of the recombination takes place on dust surface. Dust surface in protoplanetary disk absorb charged and neutral gas particles, and act as a catalyst in the chemistry.

In short, dust aggregates in plasma are negatively charged due to electron absorption. Now the detail. One can formulate the time evolution of the charge q of one dust aggregate as follows:

$$\frac{dq}{dt} = en_i\sigma_i v_i - en_e\sigma_e v_e \quad (1)$$

where e is the elemental charge, subscript $_i$ and $_e$ denote the ion and electron respectively, n is the number density of ion or electron, σ is the absorption cross section, v is the relative collision velocity. The ion species whose charge is not $+1$ are omitted for simplicity.

The relative collision velocity is dominated by the thermal velocity of the plasma particles. Since the electron mass is lighter than molecular ion species by the factor of several thousands, v_e is of the order of $100v_i$. This cause the dust aggregate to absorb more electron than ions, which makes the dust to charge negatively. The electric repulsive force from the negatively charged dust make it harder for electrons to hit the dust, and easier for ions. i.e. σ_e becomes smaller and σ_i becomes larger. The equilibrium is reached when $\sigma_i v_i = \sigma_e v_e$. At this point, about 99% of thermal electrons heading to the dust are repulsed away.

The role of dust as plasma absorber and recombination catalyst has been well studied. However, the effect of this negative dust charge on dust coagulation has strangely been ignored until recently. Okuzumi[33] studied this effect in detail. The result seemed discouraging (Figure 4). At the very initial stage of the dust growth, the repulsive force between the negatively charged dust particles is so large that the dust collision is severely hindered and dust growth is frozen in most of the dust-rich regions of the protoplanetary disk. This is explained as follows. At the very beginning of planet formation when the dust is very small, its main velocity is Brownian motion, which have the same temperature (kinetic energy) as the plasma. Now each dust is so charged that about 99% of thermal electrons heading to the dust are repulsed away and the charge flux to the dust is in equilibrium. Therefore, another dust, when it has the few times negative charge and the same kinetic energy compared to an electron, almost cannot touch the dust. Thus dust growth by coagulation is frozen.

The dust growth is permitted only at the regions with high altitude and very long distance from the star, where the dust charge is low due to low temperature, and sedimentation speed is large (and is much faster than Brownian motion) due to low disk gas density.

The solution come out when we give up one of the physicists' favorite assumption that all dust aggregates are of equal size. The planetesimal seeds grown at the outer region of the disk eventually begin to migrate inward. Due to the large migration velocity brought by their size, the outsider seeds can overcome the electrostatic barrier and capture the dust at the growth prohibited region. Here, the size distribution of the dust is crucial. The smaller dust, major in their number, absorb the plasma and although they cannot stick together themselves, contributes in not making the electrostatic barrier too high. The larger dust from outside, gathers such smaller dust and become large planetesimals. While the motion of smaller dust is coupled to the fluid, larger dust migrate at its terminal velocity towards the center of gravity.

This situation quite resembles what happens in the rain cloud on Earth. In the cloud small ice particles are coupled to the ascending air current. At the top of the cloud larger ice grains are formed. They eventually decouple from the fluid and fall towards the central star (the Earth). The larger ice collide with smaller ice and grow even larger to become snow or rain we observe. The bimodal size distribution, their different response to the gas and the fluid resulting in mean motion between them, the process they grow, is qualitatively the same between protoplanetary disks and Earth clouds.

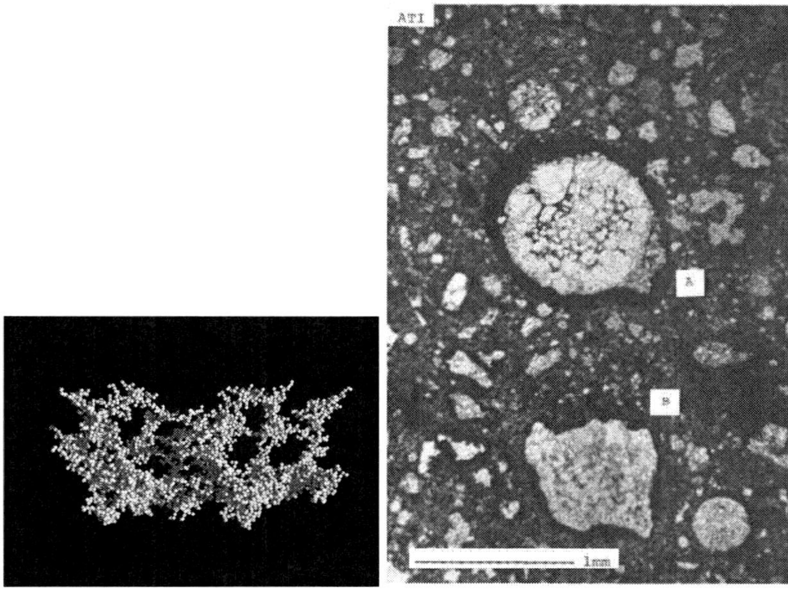

Figure 5. The dust aggregates, made of stone or ice, in a protoplanetary disk has very low-density, fluffy structure when $r < 1$ m. This is because any compacting force (e.g. self-gravity) is too weak. (Left) A dust aggregate image from structural evolution simulation, by Wada et al. [50]. The monomer, the smallest unitary sphere in the image, is supposed to be $0.1\ \mu$m in diameter. (Right) An alpha-track image of a sliced chondrite by Shaw et al. [38]. A large chondrule is marked by [A], a mm-sized, dense, spherical structure. Such chondrules are evidences for transient heating and cooling events, where fluffy dust aggregates are melted, become droplets, and frozen again to form chondrules.

An interesting aspect of ice particles is that they also exchange electric charge when they collide, which leads to lightning in certain conditions. So what happens when we import this collisional charge separation mechanism, into protoplanetary disks? The next two sections deal with this problem, according to one of our paper [30]. Please refer to the paper for more detail.

3. Dust Charge due to Collision of Ice Dust

3.1. Earlier Studies on Disk Lightning

To apply Earth lightning model in protoplanetary disk, to tell you the truth, is not a new idea. Once lightning in protoplanetary disk was studied as a candidate of chondrule heating mechanism, and has been almost rejected.

Chondrules are spherical, silicate (silicon oxides) grains of mm size, found today in wide area of our solar system (Figure 5, right.) They are usually found included in chondrule-rich meteorites, called chondrites. Their round shape and their chemical composition suggest that they are formed from fluffy silicate dust aggregates when they are subject to 1200K heat, melted, become spheric by surface tension, then frozen. The melting temperature of silicate is $10^1 \sim 10^2$ times higher than mean temperature of the disk regions considered. Also we know from the fraction of volatile materials in chondrules that the heating did not last too long. Some transient, unknown heating source must have existed.

Lightning has been a candidate mechanisms for such transient heating, although compared to other models e.g. heating by shock wave [e.g. 29], difficulties have been pointed out. Such difficulties can be summarized as three problems:

- **Neutralization Problem** [15].

  Lightning due to ice-ice dust collision is not likely to take place. In protoplanetary disk because of plasma conductivity and few dust number density, the generated electric field E is too weak. You can improve E by increasing the dust number density n, but $E \propto n^2$ so you need unnaturally high dust number density.

- **Energetics Problem** [52, e.g.].

  If lightning do take place, it is not energetic enough to melt all the chondrules observed today, because the energy budget is turbulent energy — a little portion of total kinetic energy.

- **Destruction Problem**[16].

  If lightning do take place, experiments show that it destroys the dust aggregates rather than melts them. Güttler et. al. synthesized the fluffy dust sample imitating the protoplanetary dust, placed the sample between two electrodes in a vacuum chamber, and exposed it to a electrical discharge. The dust samples generally fragmented leaving the major fraction thermally unprocessed.

So, is the protoplanetary disk lightning impossible? New knowledge in protoplanetary disk has been coming in since then, and that motivates us to reconsider the possibility. Dust aggregates are not in ordinal density — they take a very low-density, fluffy form as in Figure 5 (left), due to coagulation growth. This helps a lot in increasing dust-dust collision cross section in favor of lightning. Regions with dust number density higher than “standard scenario” are possible, due to snowline [8, e.g.], streaming instability, or turbulence. We begin by taking these into consideration and re-estimate the equilibrium charge density in the dust-plasma system.

3.2. Introduction to the Model and the Two Questions

We focus on a local, uniform box at certain orbital radius r near the equatorial plane of the protoplanetary disk (Figure 6). We model the protoplanetary disk based on the minimum-mass solar nebula (MMSN) model [17], which is accepted by the “standard scenario of planet formation.”

There are five species of particle in the model (c.f. Table 1): Large and small dust made of ice; ionized molecule(i); electrons(e); and neutral gas molecule(g). For each species **I** we aim to calculate the charge per particle $q_{\rm I}$, number density $n_{\rm I}$, and charge density carried by the species $Q_{\rm I} = q_{\rm I} n_{\rm I}$. The dust distribution is not necessary bimodal: various dust

Table 1. The classification of particles we use in this chapter. 'Particle' is generic term for all components in the protoplanetary disks. Solid components are 'dust,' and the others are 'gas.' 'Gas' components are further subdivided into 'neutral gas,' and charged components, or 'plasma.' Finally, 'plasma' consists of 'electron,' the negative charge carrier, and various molecular 'ion,' the positive charge carrier. On the right side of the table, 'dust' is classified by their size as 'smaller dust' and 'larger dust.' Either can be 'anionic' or 'cationic' dust, depending on the material they consist of. We also use the one-letter symbols 'g', 'e', 'i', 'S', and 'L' for neutral gas, electron, ion, Smaller and Larger dust. The symbols for 'Cationic' and 'Anionic' dust are 'C' and 'A'. We use variable I to represent one of these symbols.

particle						
gas				dust		
	plasma					
neutral gas	ion	electron		smaller dust		larger dust
					×	
neutral	cation	anion		cationic		anionic

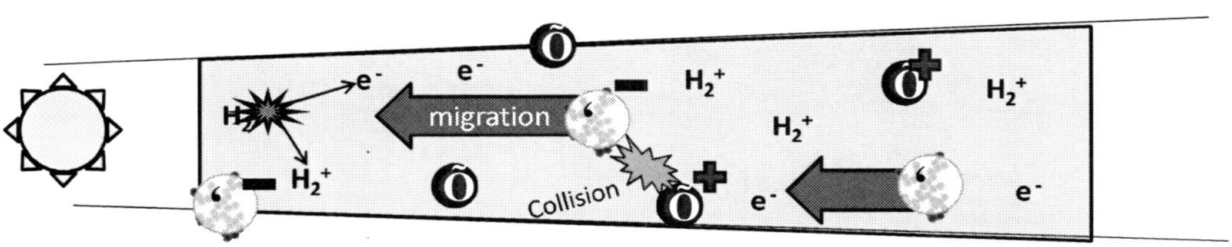

Figure 6. The uniform-box model, and the participating particles: Large dust(L); Small dust(S); ionized molecules (H_2^+); electrons(e^-); and neutral molecular gas(H_2). Neutral molecule is ionized by cosmic rays. Charged gas particles are absorbed to dust particles. Dust particles, made of ice, separate charge when they collide. The large dust species has mean velocity compared to other species.

species are classed into two groups, and the two dust (L) and (S) are representatives, or averages. Depending on which phase of the disk or which kind of the disk you consider, the parameters (e.g. radii, material density, surface condition ...) of the dust species may vary. The most focused parameter is И, the number density of the dust in units of MMSN value. Dependencies on И can either be interpreted as time dependency (as dust sedimentation or condensation proceed, И increase with time) or simply as different types of disk (some disk has more dust than other.)

The question is twofold:

1. For a certain combinations of dust parameters, how does the dust charge Q_{I}, and the generated electric field E_{esf}, depends on И? The question is the theme of this section, is detailed along with introducing the components of our model.

2. On answering the first question, we define И_{crit} so that $E(\text{И}_{\mathrm{crit}}) = E_{\mathrm{crit}}$; when we increase dust number density И while other dust parameters are fixed, the generated

electric field E_{esf} gradually rises, and $И = И_{\mathrm{crit}}$ is the point lightning strikes. Our second question is to determine $И_{\mathrm{crit}}$ as functions of dust parameters. The question is the theme of §4..

3.3. The Disk Model

The physical parameters of the protoplanetary disk borrowed from the MMSN are as follows. The gas surface density $\Sigma_g^{\bowtie}(r)$, disk scale height $h^{\bowtie}(r)$, and the temperature $T^{\bowtie}(r)$ of the disk are

$$\Sigma_g^{\bowtie}(r) = 1.7 \times 10^3 \left(\frac{r}{AU}\right)^{-\frac{3}{2}} \text{ g cm}^{-2}, \tag{2}$$

$$h^{\bowtie}(r) = 4.7 \times 10^{-2} \left(\frac{r}{AU}\right)^{\frac{5}{4}} \text{ AU}, \tag{3}$$

$$T^{\bowtie}(r) = 2.8 \times 10^2 \left(\frac{r}{AU}\right)^{-\frac{1}{2}} \text{ K}, \tag{4}$$

where r is the distance from the central star. This leads to gas density distribution

$$\rho_g^{\bowtie}(r) = 2.4 \times 10^{-9} \left(\frac{r}{AU}\right)^{-\frac{11}{4}} \text{ g cm}^{-3}. \tag{5}$$

The dust-to-gas ratio in MMSN is approximately 1.0×10^{-2}.

We use the model by Cuzzi & Zahnle[8], and introduce the two species of dust, the smaller and the larger. We further assume that the surface density of the larger dust is 10% of the total dust surface density. These two species are also either 'cationic' and 'anionic.' The 'cationic' species receives the positive electric charge through dust-dust collision. We can also represent the role of various molecular ions by one abstract ion species ' i,' according to Okuzumi[33].

The motivation for this two-dust model is twofold. First, the two dust model is the simplest model that can handle the dust-dust collisional charge separation and the macroscopic relative velocity between the dust species. Second, the charge tendency of the dust and their size are strongly correlated. In one scenario, older dust are larger and also anionic. In another scenario, dust made of ice is larger and also cationic compared to dust made of silicate. (see §3.7. for the details.) Therefore, we expect that instead of considering four (cationic smaller dust, cationic larger dust, anionic smaller dust, and anionic larger dust) species of dust, we can correlate the two size species with the two charge tendency species, (Table 1), although both correspondences (smaller dust is cationic / larger dust is cationic) are possible.

To summarize, we define the reference density of the smaller dust $\rho_S^{\bowtie}(r)$ and the density of the larger dust $\rho_L^{\bowtie}(r)$ as

$$\rho_S^{\bowtie}(r) = 1.0 \times 10^{-2} \rho_g(r), \tag{6}$$

$$\rho_L^{\bowtie}(r) = 1.0 \times 10^{-3} \rho_g(r). \tag{7}$$

We further assume that within the focused region, density for each dust component of the disk are multiplied from the MMSN by the factor $И$. Then the density of gas, smaller

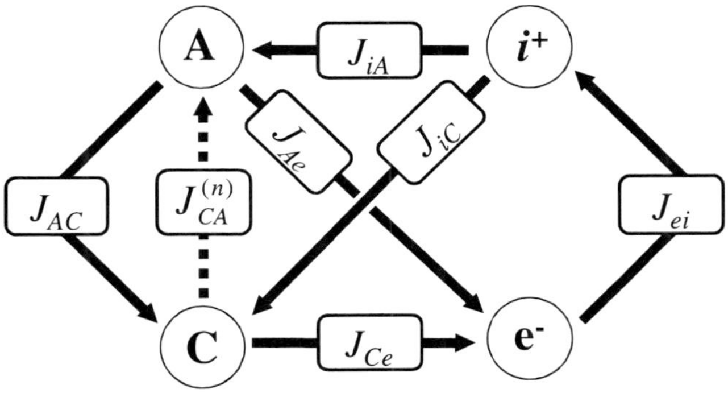

Figure 7. The circuit diagram of the charge exchange process in dust plasma. Each arrow represents 'current' density J, which has the unit $\mathrm{esu\,cm^{-3}s^{-1}}$, the amount of charge passed from one component to the other per unit disk volume per unit time. The arrow points from the component that receives negative charge to the component that receives positive charge. In this figure, i and e are ions and electrons created from ionizing neutral disk gas. C and A are cationic and anionic dust.

J_{ei} represents gas ionization as 'current' from e vertex to i vertex; J_{iA}, J_{iC}, J_{Ae}, and J_{Ce} are ion and electron absorption to dust; J_{AC} is dust-dust collisional charge separation and $J^{(n)}_{CA}$ is neutralization current of charged dust-dust absorption.

dust and larger dust is given by

$$\rho_g = \rho_g^{\bowtie}(r), \tag{8}$$
$$\rho_S = И\rho_S^{\bowtie}(r), \tag{9}$$
$$\rho_L = И\rho_L^{\bowtie}(r). \tag{10}$$

Mass of the smaller dust and the larger dust are m_S and m_L, respectively. The number density is density divided by dust mass:

$$n_S = \frac{\rho_S}{m_S}, \tag{11}$$
$$n_L = \frac{\rho_L}{m_L}. \tag{12}$$

We estimate the mass as a function of the dust radius and the fractal dimension in §3.5..

3.4. Charge Exchange Equations

There are four species of charge carrier in our model — ions, electrons, cationic, and anionic dust (Table 1.) Charge exchange processes between these species considered here are ionization, plasma absorption, and dust-dust collision/neutralization (Figure 6.)

We label the particle species with letter **I**. The charge density carried by species **I** is Q_{I} (the unit is $\mathrm{esu\ cm^{-3}}$), and the charge transferred from species **I** to species **I′** is $J_{\mathrm{I,I'}}$ (the unit is $\mathrm{esu\ cm^{-3}\ s^{-1}}$).

The charge density Q_{I} of a species **I** is the product of their number density n_{I} and their average charge per particle q_{I}. For dust species, we assume that n_{I} is known from number density model while q_{I} is unknown; for ion and electrons we know q_{I} but do not know n_{I}. This constitutes the four dynamical equations for four unknown variables q_A, q_C, n_i, n_e :

$$\frac{dq_A}{dt} = \frac{1}{n_A}\left(-J_{A,C} + J_{i,A} - J_{A,e} + J^{(n)}_{C,A}\right), \tag{13}$$

$$\frac{dq_C}{dt} = \frac{1}{n_C}\left(J_{A,C} + J_{i,C} - J_{C,e} - J^{(n)}_{C,A}\right) \tag{14}$$

$$\frac{dn_i}{dt} = \frac{1}{q_i}\left(J_{e,i} - J_{i,A} - J_{i,C}\right) \tag{15}$$

$$\frac{dn_e}{dt} = \frac{1}{q_e}\left(-J_{e,i} + J_{A,e} + J_{C,e}\right). \tag{16}$$

The current terms $J_{\text{I},\text{I}'}$ are

$$J_{A,C} = \Delta q_{A,C} n_A n_C \sigma_{A,C} \Delta v_{A,C} \tag{17}$$

$$J^{(n)}_{C,A} = |q_S|\, n_A n_C \sigma^{(n)}_{A,C} \Delta v_{A,C} \tag{18}$$

$$J_{i,A} = e n_i n_A \sigma_{cou}\left(q_A, e\right) v_i \tag{19}$$

$$J_{i,C} = e n_i n_C \sigma_{cou}\left(q_C, e\right) v_i \tag{20}$$

$$J_{A,e} = e n_e n_A \sigma_{cou}\left(q_A, -e\right) v_e \tag{21}$$

$$J_{C,e} = e n_e n_C \sigma_{cou}\left(q_C, -e\right) v_e \tag{22}$$

$$J_{e,i} = \zeta n_g \tag{23}$$

where we have included neutral gas ionization $J_{e,i}$, dust-plasma absorption $J_{A,i}$, $J_{C,i}$, $J_{A,e}$, $J_{C,e}$, dust-dust collisional charge-up $J_{A,C}$, and dust-dust collisional neutralization $J^{(n)}_{C,A}$ terms. Here, v_i and v_e are the thermal velocity of the ions and the electrons, n_g is the number density of the neutral gas, ζ is the ionization rate, which is dominated by cosmic ray ionization near equatorial, $r = 2.7$ AU of MMSN [49].

We want to solve the equilibrium equations for the dynamic equations (13-16):

$$-J_{A,C} + J_{i,A} - J_{A,e} + J^{(n)}_{C,A} = 0 \tag{24}$$

$$J_{A,C} + J_{i,C} - J_{C,e} - J^{(n)}_{C,A} = 0 \tag{25}$$

$$J_{e,i} - J_{i,A} - J_{i,C} = 0 \tag{26}$$

$$-J_{e,i} + J_{A,e} + J_{C,e} = 0, \tag{27}$$

together with charge neutrality equation:

$$Q_A + Q_C + Q_i + Q_e = 0. \tag{28}$$

We use circuit diagram (Figure 7) to depict the dynamical equations (13-16), and to interpret the numerical equilibrium solutions (24-28) in §3.10.. The circuit diagram represents charge-exchange processes; each vertex represents the species of charge reservoir and each arrow represents the charge exchange process. The size of the vertex circles represents the amount of charge Q_{I}. The thickness of the arrows represents the amount of charge transfer $J_{\mathrm{I},\mathrm{I}'}$. We define the direction of the arrows so that the arrows point to the positive charge receivers.

In the system of equations depicted by a circuit diagram, charge density of each vertex Q_{I} corresponds to an unknown quantities. Therefore, the number of unknown quantities is equal to the number of vertexes N_V. On the other hand, at the equilibrium, sum of the current flowing into each vertex is required to be zero (Kirchhoff's Laws); this gives us N_V equations but only $N_V - 1$ of them are independent. Charge neutrality gives us 1 equation. Thus we have N_V equations for N_V unknown values.

3.5. Fluffy Dust Model

We use model of dust aggregates by Wada et al.[50]. c.f. Figure 5. We consider dust aggregates composed of a large number of spherical monomers with radius $r_m = 0.1\ \mu\mathrm{m}$. Each dust species **I** has its mass m_{I}, the number of monomers that constitute the dust N_{I}, and representative radius r_{I}. We define the fractal dimension of the fluffy dust D_{I} in the following simple manner:

$$N_{\mathrm{I}} = \left(\frac{r_{\mathrm{I}}}{r_m}\right)^{D_{\mathrm{I}}} . \tag{29}$$

The dust mass is expressed in terms of monomer mass m_m as follows:

$$m_{\mathrm{I}} = m_m N_{\mathrm{I}} = m_m \left(\frac{r_{\mathrm{I}}}{r_m}\right)^{D_{\mathrm{I}}} . \tag{30}$$

Wada et al.[50] studies the collision of the fluffy dust of the radii $1.0 \times 10^{-5} \sim 9.1 \times 10^{-4}$ cm. The effect of offset collisions, collision between dust of much different sizes, and dust much larger than 9.1×10^{-4} cm are yet to be confirmed. Therefore we make the following assumptions on smaller dust-larger dust collision.

- If the smaller dust graze at the larger dust, i.e. if the line that passes the gravitational center of the smaller dust and is parallel to the relative velocity vector do not intersect with the larger dust, the two dust aggregates do not stick to each other. Therefore the grazing cross section is of the order of $r_{_S} r_{_L}$. In this case they separate $\Delta q_{A,C}$ of charge, which is the product of charge surface density σ_{ch} and contact surface area S_{kiss}. This contributes to the dust-dust charging current, $J_{C,A}$.

- If the smaller dust bump into the larger dust, i.e. if the line that passes the gravitational center of the smaller dust and is parallel to the relative velocity vector do intersect with the larger dust, the smaller dust do not penetrate the larger dust but becomes a part of the larger dust. The cross section is of the order $r_{_L}{}^2$. In this case all the charges the smaller dust have are removed from the smaller dust charge density and added up to the larger dust charge density. This contributes to the dust-dust neutralization current, $J^{(n)}_{C,A}$.

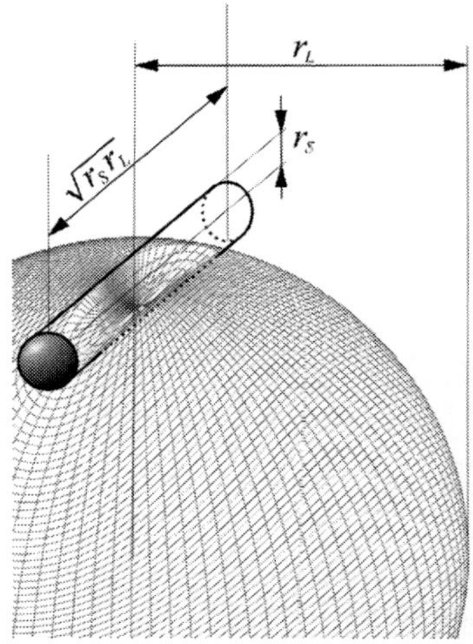

Figure 8. A grazing collision between a smaller dust (the blue solid sphere) and a larger dust (the red wire-frame sphere). The smaller dust creates a trench on the larger dust (the black cylinder). Be noted that two spheres are not solidly filled; they are filled with μm-sized monomers and there are lots of space.

3.6. Collisional Cross Section and Contact Surface of Charged Fluffy Dust

In this section, we estimate collisional cross sections for dust. The collisional cross sections for two electrically charged spherical particle is given by

$$\sigma_{cou}(q) = \pi a^2 \exp\left(-\frac{qq'}{ak_BT}\right) \qquad (qq' > 0), \tag{31}$$

$$\sigma_{cou}(q) = \pi a^2 \left(1 - \frac{qq'}{ak_BT}\right) \qquad (qq' < 0) \tag{32}$$

where q,q' is each particle's charge, T is the temperature of their relative motion and πa^2 is the geometric cross section [e.g. 42].

Equation (32) represents the effect of Coulomb focusing: particles of the opposite charge attract each other and collide more often than when they are neutral. On the limit $|qq'a^{-1}| \gg k_BT$ we can approximate the cross section as $\sigma_{cou}(q) \simeq -\pi aqq'(k_BT)^{-1}$, which is bi-linear on q and q'. On the other hand, cross section (31) represents the effect of Coulomb repulsion: for the collision between particles of the same charge only a portion of particles that belongs to the long tail of Boltzmann's distribution for temperature T can overcome the Coulomb barrier and collide. On the limit $qq'a^{-1} \gg k_BT$ the cross section vanishes quickly, but never reaches 0.

We use this Coulomb cross sections (31), (32) to estimate the event rate of gas-dust collision and dust-dust collision.

The amount of charge exchanged in a collision, $\Delta q_{A,C}$, is product of area of contact S_{kiss}, upper limit of charge exchanged per unit surface area of contact σ_{ch}, and the non-dimensional efficiency factor $И_{ch}$. We leave the detailed argument to determine $И_{ch}\sigma_{ch}$ to

§3.7.. Here we assume that $И_{ch}\sigma_{ch}$ is known and describe how to estimate contact surface area $S_{\rm kiss}$. Since it requires another detailed simulation to estimate $S_{\rm kiss}$ qualitatively, we resort to an order-of-magnitude estimate for this part of the work.

We illustrate the collision between a smaller dust and a larger dust in Figure 8. The smaller dust grazes the larger dust, pushes away the monomers that belong to the larger dust and creates a trench on the larger dust. The trench is a portion of the black cylinder in the figure. The radius and the length of the cylinder is r_S and $(r_S r_L)^{1/2}$, respectively. Therefore, the surface area of the trench S_C is of order

$$S_C \simeq {r_S}^{3/2} {r_L}^{1/2}, \tag{33}$$

and the number of monomers N_C required to fill the surface of the trench is

$$N_C \simeq {r_m}^{-2} {r_S}^{3/2} {r_L}^{1/2}. \tag{34}$$

Their total surface area is also of the order of S_C.

However, $S_{\rm kiss} \simeq {r_S}^{3/2} {r_L}^{1/2}$ overestimates the actual contact surface area if the large dust is so fluffy that there is not enough monomers in the trenched volume to fill the trench surface.

From the definition of the fractal dimension (29), the number density of monomers within the larger dust material is

$$n_L^{(M)} = N {r_L}^{-3} = {r_m}^{-D_L} {r_L}^{D_L - 3}. \tag{35}$$

On the other hand the volume of the trench is

$$V_F \simeq {r_S}^{5/2} {r_L}^{1/2}. \tag{36}$$

Therfore, the number of particle contained in the trench is

$$N_F = n_L^{(M)} V_F \simeq {r_m}^{-D_L} {r_S}^{5/2} {r_L}^{D_L - 5/2}, \tag{37}$$

and their total surface area is

$$S_F \simeq {r_m}^2 N_F \simeq {r_m}^{2-D_L} {r_S}^{5/2} {r_L}^{D_L - 5/2}. \tag{38}$$

If $N_F < N_C$, the surface of the trench is only partially covered by the monomers, and we estimate $S_{\rm kiss} \simeq S_F$. On the other hand, if $N_F > N_C$, N_F monomers are crushed onto the trenched surface, and since they overlap, about N_C monomers will take part in the charge exchange. In this case we estimate $S_{\rm kiss} \simeq S_C$. To summarize, we assume that $S_{\rm kiss}$ is the smaller of (33) or (38):

$$S_{\rm kiss} = \min\left({r_S}^{3/2} {r_L}^{1/2}, {r_m}^{2-D_L} {r_S}^{5/2} {r_L}^{D_L - 5/2} \right). \tag{39}$$

3.7. Charge Separation Processes

There are generally two classes of possible charge separation processes in protoplanetary disks.

One is surface charge exchange, where each dust has some kind of spontaneous charge separation [23], so at the initial condition each dust charge is zero as a whole (globally neutral), but there are charge separation within the dust particles (locally charged). For example, water ice crystals tend to gather negative charge at its surface and positive charge inside. When two dust aggregates with different charge collide and melt partially, they exchange molten material and the charge included in the molten material. As a result each dust gets globally charged.

The other charge separation mechanism may be triboelectric processes [e.g. 10]. In this case, at the initial condition each dust is both globally and locally neutral. When two dust aggregates made of materials with different electron affinity collide, the surface electrons move from one material to the other. As a result each dust gets globally charged.

3.7.1. Surface Charge Exchange I — Larger Dust Is Anionic

The mechanism we consider the most plausible for the dust-dust collisional charge separation is surface charge exchange between ice dust. For the dust aggregate of ice mantled silicate, Cuzzi & Zahnle[8] proposed a condensation scenario, that at the snow line ice larger dust drifting inward dissociate and many smaller dust form.

There are established models on charge separation caused by ice-ice dust collision in the context of thundercloud meteorology (we referred to, e.g. Dash et al.[9]). We will carefully import them as a charge separation model in protoplanetary disks. The essential steps to cause lightning on earth are (1) spontaneous charge separations on ice crystal surfaces, (2) existence of different dust species with different spontaneous charge separation per surface area, (3) collisions between the different dust that leads to global charging of each dust and (4) relative motion between the globally charged dust to create electrostatic field.

For (1), we argue that the charge separation per surface area is quantitatively the same as the values measured in laboratory experiments. For (2), dominating dust species in charge separation process in protoplanetary disks is uncertain, and we discuss two possibilities (c.f. §3.7.1., §3.7.2.) in this work. For (3) and (4), we make simple estimations for the collision rate and relative velocity in protoplanetary disks.

Ice crystal surface is intrinsically charge-separated. Ice is negatively charged near the surface, and the inside is positive. The typical charge surface density for stable ice surface is $\sigma_{ch} \simeq 3.0$ esu cm^{-2} or $\sigma_{ch} \simeq 6.2 \times 10^9$ e cm^{-2} and the typical skin depth of the charged layer is $d_{ch} \simeq 2.0 \times 10^{-4}$ cm, though charge surface density for fast-growing ice surfaces are larger and shallower [9]. This charge separation has a general explanation as a result of interaction between hydroxide(OH^-) and hydronium (H_3O^+) ions and a hydrophobic surface [23, 31], and the above value of typical charge surface density is observed at liquid water-air surfaces as well as at ice crystal-air surfaces [45]. Therefore we use the value for ice-vacuum surfaces as well.

In the thundercloud, there are varieties of ice crystals with different surface charge densities, depending on the surface history of the ice crystals. Newly formed surfaces have larger charge surface density than old surfaces, because they have higher fractal dimension and deeper amorphous layers.

We now consider how surface charge exchange works in the model of Cuzzi & Zahnle [8]. Larger dust that migrate towards the snow line has old surface and has less negative charge surface density, while smaller dust formed at the snow line have new surface and larger negative charge surface density, as in meteorological case. Note that before collision each dust is globally neutral.

At the collision, the surface of the dust aggregates melts and the surface charge density is exchanged, and averaged. The larger dust, having less surface charge density than the smaller dust, receives more negative charge than it gives. Therefore the larger dust becomes anionic, smaller dust becomes cationic.

Laboratory experiments [46], in-situ observations and meteorological estimates [14, 7] suggest that for mm-size ice crystals, at least 10 per cent of the total surface charge within contact surface is exchanged in a single collision; experiments by Mason & Dash[28], Dash et al.[9] suggests almost $И_{ch} = 1.0$. As a conservative estimate, we use $И_{ch} = 0.1$ unless mentioned otherwise.

3.7.2. Surface Charge Exchange II — Larger Dust Is Cationic

It may be possible that charge separation processes occurring in protoplanetary disks are different from those occurring in the terrestrial thunderclouds. The collision time-scale in the protoplanetary disks is much longer than that in a thundercloud, so long that sintering may take place [39]. As a result, The surface state of old ice larger dust and young ice smaller dust might resemble each other. If they are identical, some random charge exchange by collision is still possible, but they do not exchange charge on average.

However, compared to thundercloud, protoplanetary disks are more dirty and fine-grained; they contain much dust made of materials other than ice such as silicates, and the monomer size is $0.1\ \mu\mathrm{m}$ rather than $1\ \mathrm{mm}$. Since the monomer size is smaller than typical skin depth of the charge separation $d_{ch} \simeq 2.0 \times 10^{-4}$ cm mentioned above, it is possible that ice smaller dust and silicate smaller dust with thin ice mantles formed at the snow line is inefficient in separating charge. There may be silicate aggregates with no surface charge separation. Meanwhile old larger dust that have traveled from the far end of the protoplanetary disk have undergone sintering and have developed thick mantles with full surface charge separation.

In such scenario, the larger dust has more surface charge separation than the smaller dust. Therefore, collision between a larger dust and a smaller dust still leads to charge separation but the larger dust becomes cationic, and the smaller dust is anionic in this case. We assume that $И_{ch} = 0.1$ and $\sigma_{ch} \simeq -3.0$ esu cm^{-2} in this case (The charge exchange rate has the same magnitude but the opposite sign compared to that of §3.7.1..)

Both scenarios, the larger dust is anionic and the larger dust is cationic are plausible. They may even take place in the different parts of the same disk simultaneously. Therefore, we have decided to take both scenarios into consideration. To that end, we treat the concept of cationic and anionic dust separately from the size of the dust.

3.8. Relative Velocity

When a cloud of positively and negatively charged dust is separated much larger than plasma Debye length

$$\lambda_D = \sqrt{\frac{T}{4\pi n_i e^2}} = 4.0 \times 10^2 \text{ cm} \left(\frac{T}{170K}\right)^{\frac{1}{2}} \left(\frac{n_i}{5.0 \times 10^{-2} cm^{-3}}\right)^{-\frac{1}{2}} \tag{40}$$

the electrostatic field between them become observable. In order to cause such macroscopic charge separation, there must be a significant relative bulk motion between anionic and cationic dust. Inward migration of large dust is a source of this bulk motion. The sedimentation may act in the same way. Also Desch & Cuzzi[10] have proposed that largest eddies in turbulence of protoplanetary disks cause bulk motion between smaller dust and larger dust. Such effects on the relative velocity between dust species in MMSN has been studied (see Brauer et al.[5] and references therein).

Here, we simply assume that the largest contribution to the smaller dust-larger dust relative velocity is the bulk motion of the larger dust, and the velocity is $\Delta v_{L,S} \equiv u_L \equiv 3.4 \times 10^3$ cm s^{-1}, the catastrophic collision velocity of the ice dust aggregates of 9.1×10^{-4} cm size dust [50]. Note that the non-sticking velocity threshold decrease as the monomer size increase [3]. We also check our analytic formulae with smaller values of $\Delta v_{L,S}$ and u_L assumed.

Dust migration speed are comparable to this value at some stages of the dust growth. On the other hand, turbulent motion is faster than the value for most of our parameter range. Turbulent mode that is larger than the scale of interest can be treated as bulk motion, and can be used to explain the charge separations of the scale. The scale can be as large as of order of disk scale-height [1].

3.9. The Charge Equilibrium Equations

By substituting the results of analyses up to here into (13-16) we have the following dynamic equation for charge transport:

$$\frac{dQ_L}{dt} = -J_{L,S} - J_{L,i} - J_{L,e} \tag{41}$$

$$\frac{dQ_S}{dt} = J_{L,S} - J_{S,i} - J_{S,e} \tag{42}$$

$$\frac{dQ_i}{dt} = J_{L,i} + J_{S,i} - J_{i,e} \tag{43}$$

$$\frac{dQ_e}{dt} = J_{L,e} + J_{S,e} + J_{i,e} \quad , \tag{44}$$

where the current density terms (17-23) become:

$$J_{L,S} = \left(\frac{2r_S}{r_L}\Delta q_{A,C} - \frac{Q_S}{n_S}\right) n_S n_L \Delta v_{L,S}$$
$$\sigma_{cou}\left(\frac{Q_L}{n_L}, \frac{Q_S}{n_S}, r_L, \frac{1}{2}m_S {\Delta v_{L,S}}^2\right) \quad (45)$$

$$J_{L,i} = -Q_i n_L \sigma_{cou}\left(\frac{Q_L}{n_L}, e, r_L, k_B T\right) v_i \quad (46)$$

$$J_{L,e} = -Q_e n_L \sigma_{cou}\left(\frac{Q_L}{n_L}, -e, r_L, k_B T\right) v_e \quad (47)$$

$$J_{S,i} = -Q_i n_S \sigma_{cou}\left(\frac{Q_S}{n_S}, e, r_S, k_B T\right) v_i \quad (48)$$

$$J_{S,e} = -Q_e n_S \sigma_{cou}\left(\frac{Q_S}{n_S}, -e, r_S, k_B T\right) v_e \quad (49)$$

$$J_{i,e} = -e\zeta n_g. \quad (50)$$

In (45), the amount of current exchange $\Delta q_{A,C}$ is product of contact surface area S_{kiss} and surface charge density σ_{ch}, each described in §3.6. and §3.7.. The contact surface area S_{kiss} is the function of dust radii and dust fractal dimensions; see equation (39). The surface charge density σ_{ch} depends on the dust material. The relative velocity of the larger dust and the smaller dust is $\Delta v_{L,S} = 3.4 \times 10^3$ cms^{-1}, as we have discussed in §3.8.. The cross section term σ_{cou} is the Coulomb cross section introduced in §3.6.. We assume v_i and v_e to be thermal velocities of ions and electrons. For ionization in MMSN at $r = 2.7$ AU, cosmic ray ionization is the main contributor and $\zeta \simeq 10^{-18}$ [49]. We introduce the non-dimensional dust number density $И$ (dust number density in unit of MMSN values), so that in equations (8-10), $И_g = 1$, and $И_S = И_L = И$. From those density term, the number density terms n_g, n_S, n_L are given as $\rho_g/m_g, \rho_S/m_S, \rho_L/m_L$. The masses of dust aggregates m_S, m_L are function of their radii and fractal dimensions; see equation (30).

All the variables that appear in the current density terms (45-50) are controlled by five parameters; radii of the dust aggregates (r_S, r_L), their fractal dimension (D_S, D_L), and the non-dimensional dust number density $И$.

The equilibrium equations (24-28) become:

$$-J_{L,S} - J_{L,i} - J_{L,e} = 0 \quad (51)$$
$$J_{L,S} - J_{S,i} - J_{S,e} = 0 \quad (52)$$
$$J_{L,i} + J_{S,i} - J_{i,e} = 0 \quad (53)$$
$$J_{L,e} + J_{S,e} + J_{i,e} = 0 \quad (54)$$
$$Q_L + Q_S + Q_i + Q_e = 0. \quad (55)$$

Again note that, out of four Kirchhoff's Laws (51-54) only three of them are independent, and the charge neutrality condition (55) is necessary.

3.10. Results — Four Phases of Charge Separation

We found that as we increase $И$ while keeping other dust parameters constant, the equilibrium charge densities $Q_I = q_I n_I$ experience four phases (Table 2). Figure 9 and Figure 10 shows the typical four phases behavior.

In this and following sections, we explain the origin of the four phases, using the circuit diagrams (Figure 10) as a great help. The four-phase behavior we describe here is independent of most of the details of charge exchange processes. For example, when we set the opposite sign for dust-dust collisional charge exchange (§3.7.1. v.s. §3.7.2.,) the evolutions are almost similar. The rest of the discussion in following sections is based on the §3.7.1. case. The discussion is easily generalized to the other case.

To analyze the result, we first identify the dominant processes by comparing the competitive current in circuit diagram, then write down all the unknown values in simple polynomials of $И$. Figure 10 illustrates the transition of dominant process in the circuit as dust number density $И$ increases. The two particles with the largest charge density is marked by larger circle. There are always two of them, one carrying most of the system's positive charge and the other negative, thus charge neutrality holds. The arrows and their line width represents direction and amount of currents. Labels for dominant currents are marked with thick rectangle, sub-dominant currents with thin rectangle, negligible currents with dashed rectangle. The names and conditions for each phase is listed in Table 2.

There are two major consequences of the size difference. Larger dust is much fewer in number density. So in the fewer dust limit ($И \ll 1$) the larger dust carries much less charge density than smaller dust do. Since larger dust is the fewer, one larger dust collides with smaller dust much more often than one smaller dust does with larger dust. Therefore larger dust are the species that experience the quick charge density raise in (c)charge-up phase. The main role of the smaller dust is to absorb plasma and keep the charge neutrality.

Table 2. The names and conditions for four phases of charge separation. They are basically named after dominant charge carrier of each phase.

(a)	ion-electron plasma phase	$\lvert Q_e \rvert \simeq Q_i$ (in this paper $2\lvert Q_e \rvert > Q_i$)
(b)	ion-dust plasma phase	$2\lvert Q_e \rvert < Q_i$, $\lvert J_{A,C} \rvert < \lvert J_{A,e} \rvert$
(c)	charge-up phase	$\lvert J_{A,e} \rvert < \lvert J_{A,C} \rvert < \lvert J_{i,C} \rvert$
(d)	dust phase	$\lvert J_{i,C} \rvert < \lvert J_{A,C} \rvert$

3.10.1. Ion-electron Plasma Phase

In ion-electron plasma phase (Figure 10 (a)), the dominant path of charge transfer is

$$e^- \to i^+ \to C \to e^-, \tag{56}$$

the next-dominant path is

$$i^+ \to A \to e^-. \tag{57}$$

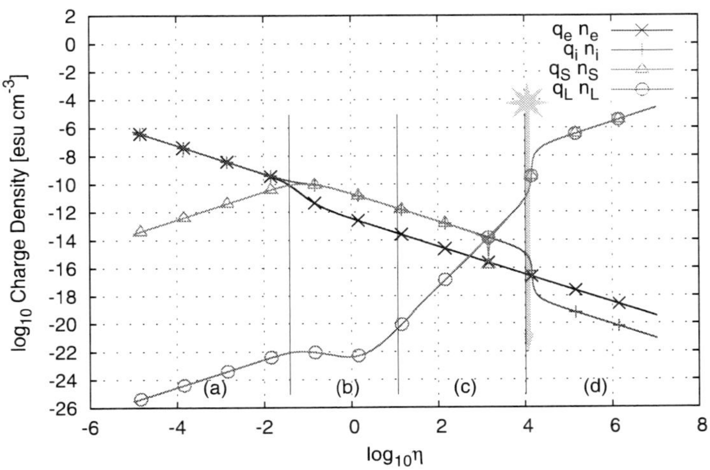

Figure 9. Amount of charge stored in each species, en_e, en_i, $|q_S|\,n_S$, and $|q_L|\,n_L$, as functions of $И$. This figure is for ice dust-ice dust case, so larger dust is anionic and smaller dust is cationic. The polarity matches that of Figure 10. The radius of smaller dust, radius of larger dust, fractal dimension of smaller dust, fractal dimension of larger dust are 1.0×10^{-4} cm, 1.0×10^{2} cm, 3.0, and 3.0 respectively. (a), (b), (c), and (d) corresponds to the four phases described in §3.10.. The yellow arrow denotes the critical number density $И$ where the macroscopic electric discharge condition (84) is met. The settings of the simulation that produces this figure is in the Table. 3.

Therefore, we have following current hierarchy:

$$\begin{aligned} J_{e,i} \simeq J_{i,C} \simeq J_{C,e} &\quad \propto И^0 \\ \gg J_{i,A} \simeq J_{A,e} & \\ \gg J_{A,C}. & \end{aligned} \tag{58}$$

The amount of current for path (56) is constrained by edge $e^- \to i^+$; since we have assumed that ζ and n_g is independent of $И$, so is $J_{e,i}$.

From charge neutrality (28), $Q_e = Q_i$ and therefore $n_e = n_i$. So equation $J_{i,C} \simeq J_{C,e}$ is satisfied by setting, in equations (20) and (22),

$$\sigma_{cou}\left(q_C, e\right) v_i = \sigma_{cou}\left(q_C, -e\right) v_e, \tag{59}$$

$$\frac{\sigma_{cou}\left(q_C, e\right)}{\sigma_{cou}\left(q_C, -e\right)} \simeq \frac{v_e}{v_i} \propto И^0. \tag{60}$$

Equation (60) tells us that $\sigma_{cou}(q_C, e)/\sigma_{cou}(q_C, -e)$ is constant of $И$. This means $q_C \propto И^0$ because the only $И$-dependent term in σ_{cou} is q_C. By definition of dust number density factor $И$, $n_C \propto И^1$, so $Q_C \propto И^1$.

By similar argument we can deduce $Q_A \propto И^1$ from $J_{i,A} \simeq J_{A,e}$.

In other hand, to satisfy $J_{i,C} \propto И^0$ and $\simeq J_{C,e} \propto И^0$ we need $n_i, n_e \propto И^{-1}$. And since $q_i, q_e \propto И^0$, we have $Q_i, Q_e \propto И^{-1}$.

In this phase, ions and electrons are the major carriers of positive and negative charge. Equation (60) also tells us that $\sigma_{cou}(q_C, e)/\sigma_{cou}(q_C, -e) \simeq v_e/v_i \gg 1$. This is interpreted

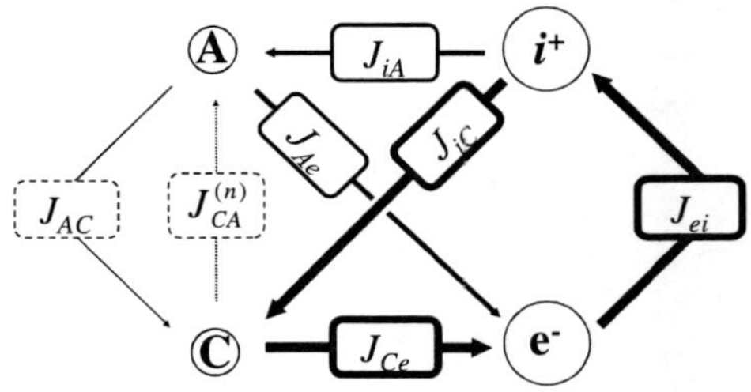

(a) ion-electron plasma phase

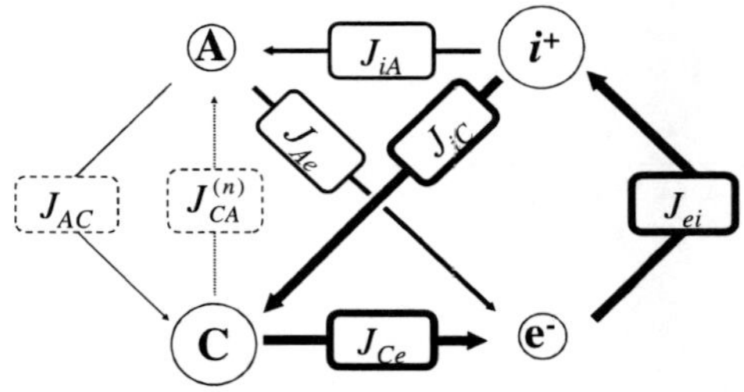

(b) ion-dust plasma phase

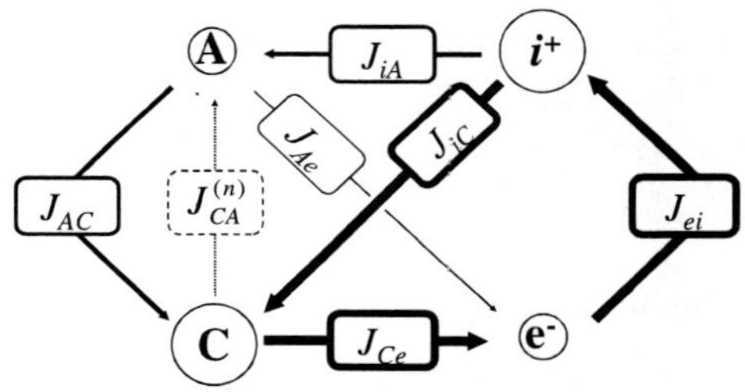

(c) charge-up phase

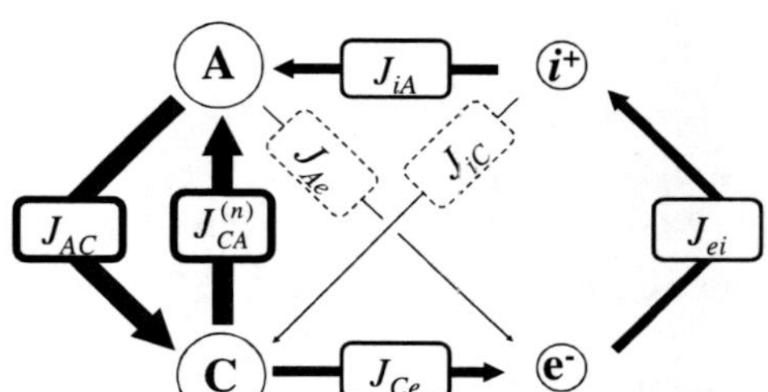

(d) dust phase

Figure 10. The evolution of the charge density and current density as $\mathit{И}$ increase. As dust number density $\mathit{И}$ increase, $J_{e,i} \propto \mathit{И}^0$ is constant while $J_{A,C} \propto \mathit{И}^2$ grows, and the particle experience four phases in order $(a) \to (b) \to (c) \to (d)$. (a) At ion-electron plasma phase, most of the charge is carried by plasma species and the charge state of the dust is governed by plasma absorption. (b) At ion-dust plasma phase, the current balances are same as it was in ion-electron plasma phase, but now the negative charge carrier is cationic smaller dust. (c) At charge-up phase, anionic larger dust has sufficient charge to cut off $J_{A,e}$. (d) At dust phase, most of the charge is carried by dust species and the collisional charging current $J_{A,C}$ balance with neutralization current $J^{(n)}_{C,A}$.

Table 3.

symbol	value
r	2.7 AU
$\Sigma_g^{\bowtie}$	3.8×10^2 g cm^{-3}
$h^{\bowtie}$	1.6×10^{-1} AU
$T^{\bowtie}$	1.7×10^2 K
$\rho_g^{\bowtie}$	1.6×10^{-10} g cm^{-3}
$\rho_S^{\bowtie}$	1.6×10^{-12} g cm^{-3}
$\rho_L^{\bowtie}$	1.6×10^{-13} g cm^{-3}
ρ_m	9.3×10^{-1} g cm^{-3}
r_m	1.0×10^{-5} cm
r_S	1.0×10^{-4} cm
r_L	1.0×10^{2} cm
D_S	3.0
D_L	3.0
ζ	1.0×10^{-18}
$\Delta v_{L,S}$	3.4×10^{3} cm s^{-1}
u_L	3.4×10^{3} cm s^{-1}
σ_{ch}	6.2×10^{9} e cm^{-2}
И_{ch}	1.0×10^{-1}

as follows: Since thermal velocity of electron is much faster than that of molecular ions, electron is more rapidly absorbed to neutral dust than ions. Therefore dust continues to acquire negative charge, until its negative charge is enough to repulse most of the electrons inflow to attain a current equilibrium. Both cationic and anionic dust are forced to charge negative to hold back the overwhelming electron absorption.

To summarize,

$$Q_i \propto \text{И}^{-1},\ Q_e \propto \text{И}^{-1},\ Q_A \propto \text{И}^{+1},\ Q_C \propto \text{И}^{+1}. \tag{61}$$

3.10.2. Ion-dust Plasma Phase

The system enters ion-dust plasma phase when the negative charge in dust Q_C become comparable to that in plasma Q_e. Charge neutrality (28) requires free electrons to decrease. So the Coulomb barrier of dust species become weaker until Coulomb cross section approximates geometric cross section $\sigma_{cou}(q_C, e) \simeq \sigma_{cou}(q_C, e) \simeq \pi {a_C}^2 \propto \text{И}^0$ where electrons and ions are equally absorbed to the dust.

In ion-dust plasma phase (Figure 10 (b)), the dominant path is still

$$e^- \to i^+ \to C \to e^-, \tag{62}$$

and the next-dominant path is still

$$i^+ \to A \to e^-, \tag{63}$$

and the same current hierarchy holds:

$$\begin{aligned} & J_{e,i} \simeq J_{i,C} \simeq J_{C,e} \quad \propto И^0 \\ \gg \; & J_{i,A} \simeq J_{A,e} \\ \gg \; & J_{A,C}. \end{aligned} \tag{64}$$

However, now that $\sigma_{cou}(q_C, e) \simeq \sigma_{cou}(q_C, e)$, equation $J_{i,C} \simeq J_{C,e}$ is satisfied by setting, in equations (20) and (22),

$$n_i v_i = n_e v_e, \tag{65}$$

$$\frac{n_i}{n_e} \simeq \frac{v_e}{v_i} \propto И^0. \tag{66}$$

So the ratio n_i/n_e is kept constant to $v_e/v_i = 6.1 \times 10^1$. Still, in order to have $J_{i,C} \propto И^0$ and $\simeq J_{C,e} \propto И^0$ we need $n_i, n_e \propto И^{-1}$. Since $q_i, q_e \propto И^0$, we have $Q_i, Q_e \propto И^{-1}$.

In this phase the cationic dust carry most of the negative charge while ions carry most of the positive charge of the system. Therefore, the charge neutrality equation (28) is dominated by these two components, and $Q_C \propto Q_i \propto И^{-1}$.

In this phase anionic dust also feels the same environment as cationic dust, so $Q_A \propto И^{-1}$. However as $И$ approaches to (c)charge-up phase, dust-dust collisional charge separation $J_{A,C}$ gradually comes into play and Q_A increases. Therefore in Figure 9 we can see the power law $Q_A \propto И^{-1}$ only at the beginning of (b)ion-dust plasma phase.

To summarize,

$$Q_i \propto И^{-1},\; Q_e \propto И^{-1},\; Q_A \propto И^{-1},\; Q_C \propto И^{-1}. \tag{67}$$

In ion-electron plasma phase and ion-dust plasma phase the dust-dust collisional charging is ineffective. So we can understand these two phase without dust-dust collisional charging (see **(author?)** [33] and references therein.)

3.10.3. Charge-up Phase

The system enters (c)charge-up phase when $J_{A,C}$ becomes larger than $J_{A,e}$. Now anionic dust has their own negative charge supply from dust-dust collision, their negative charge grow quickly, and $\sigma_{cou}(q_A, -e)$ become rapidly small. At this point, the circuit switches one of its current path.

In charge-up phase (Figure 10 (c)), the dominant path is still

$$e^- \to i^+ \to C \to e^-, \tag{68}$$

but the next-dominant path is

$$i^+ \to A \to C. \tag{69}$$

The amount of current for path (68) is constrained by edge $e^- \to i^+$; since we have assumed that ζ and n_g is independent of $И$, so is $J_{e,i}$. The amount of current for path (69) is constrained by edge $A \to C$ (17); since we have assumed that $\Delta q_{A,C}$ is independent of $И$, $J_{A,C} \propto И^2$.

Therefore, we have following hierarchy:

$$\begin{aligned} & J_{e,i} \simeq J_{i,C} \simeq J_{C,e} && \propto И^0 \\ \gg\ & J_{i,A} \simeq J_{A,C} && \propto И^2 \\ \gg\ & J_{A,e}. \end{aligned} \tag{70}$$

The path (68) is as same in ion-dust plasma phase, leading to $Q_i, Q_e \propto И^{-1}$, and charge neutrality requires $Q_C \propto И^{-1}$.

In dust charge-up phase, however, anionic dust has so much charge that electrostatic potential for electron and ion at the surface of larger dust is larger than their thermal energy; this is $qq'a^{-1} \gg k_B T$ limit of the Coulomb cross section (31), (31). Thus $\sigma_{cou}(q_A, -e) \to 0$ and $\sigma_{cou}(q_A, e) \propto q_A$ in (19). Substituting $n_i \propto И^{-1}$ and $n_A \propto И^1$ into $J_{i,A} \propto И^2$, we have $q_A \propto И^2$ and $Q_A \propto И^3$.

To summarize,

$$Q_i \propto И^{-1},\ Q_e \propto И^{-1},\ Q_A \propto И^{+3},\ Q_C \propto И^{-1}. \tag{71}$$

3.10.4. Dust Phase

The system enters (d)dust phase when $J_{A,C}$ becomes larger than $J_{i,C}$. Now the charge states of both anionic and cationic dust is governed by dust-dust collision, and the plasma component is sub-dominant to the dust.

In dust phase (Figure 10 (d)), the dominant path is

$$A \to C \to A, \tag{72}$$

the dust-dust collision is now short-circuiting. The next-dominant path is

$$C \to e^- \to i^+ \to A. \tag{73}$$

The amount of current for path (72) is constrained by edge $A \to C$ (17); since we have assumed that $\Delta q_{A,C}$ is independent of $И$, $J_{A,C} \propto И^2$.

The amount of current for path (73) is constrained by edge $e^- \to i^+$; since we have assumed that ζ and n_g is independent of $И$, so is $J_{e,i}$.

Therefore, we have following hierarchy:

$$\begin{aligned} & J_{A,C} \simeq J^{(n)}_{C,A} && \propto И^2 \\ \gg\ & J_{C,e} \simeq J_{e,i} \simeq J_{i,A} && \propto И^0 \\ \gg\ & J_{A,e}. \end{aligned} \tag{74}$$

Equation $J_{A,C} \simeq J^{(n)}_{C,A}$ (74) requires $\Delta q_{A,C}\sigma_{A,C} = q_C \sigma^{(n)}_{A,C}$. Therefore only $И$ dependent term q_C must satisfy $q_C \propto И^0$, leading to $Q_C \propto И^1$. Charge neutrality leads to $Q_A \propto И^1$.

The path (73) gives us $Q_i, Q_e \propto И^{-1}$, same as in ion-dust plasma phase and in dust charge-up phase.

At the boundary of (c)charge-up phase and (d)dust phase there is a jump of dust charge. This is because when $И$ cross the boundary dust charge grows until dust-dust collisional neutralization can compensate dust-dust charge separation.

To summarize,

$$Q_i \propto И^{-1},\ Q_e \propto И^{-1},\ Q_A \propto И^{+1},\ Q_C \propto И^{+1}. \tag{75}$$

3.11. Conclusion of the Section — Plasma with Charge-Separating Ice Dust

We have shown that as dust number density И increase, the charge density distribution experience four phases: (a)ion-electron plasma phase, (b)ion-dust plasma phase, (c)charge-up phase and (d)dust phase. The former two phases are studied in detail by Okuzumi[33], while the latter two phases are unique results of taking dust-dust collision into consideration.

4. Protoplanetary Disk Lightning

4.1. Critical Dust Number Density for Lightning

In this section we derive the strength of electric field generated by the relative motion of the large and small dust, and set conditions for macroscopic electric discharge events, or lightning.

Lightning occurs when the maximum electric field in the plasma $E_{\rm esf}$ exceeds the critical value $E_{\rm crit}$. The critical electric field $E_{\rm crit}$ is determined by the condition that an electron accelerated by the field has kinetic energy large enough to ionize a neutral gas molecule. Let $l_{\rm mfp}$ be the mean free path for electron. Then an electron accelerated in electric field of strength E receive the energy of order $e\,E\,l_{\rm mfp}$. The ionization potentials $\Delta W_{\rm ion}$ for H, H_2, and He molecules are 13.6 eV, 15.4 eV, and 24.6 eV respectively [11]. We use $\Delta W_{\rm ion} = 15.4$ eV in this work. Therefore the critical value $E_{\rm crit}$ of electric field for the lightning satisfies:

$$e\,E_{\rm crit}\,l_{\rm mfp} = \Delta W_{\rm ion}, \tag{76}$$

$$E_{\rm crit} = \frac{\Delta W_{\rm ion}}{e\,l_{\rm mfp}}. \tag{77}$$

Next we derive the value of $E_{\rm esf}$. When the differential motion between the oppositely charged dust species continues much longer than the plasma Debye length, it can be interpreted as current carried by the dust j_D generating electrostatic field, and the plasma counter-current j_p is induced in the neutralizing direction . We consider that j_p is carried by electrons, and neglect current carried by positive ions because it is at most the same order as that by electrons. Moreover, even if positive ions are accelerated to $\Delta W_{\rm ion}$ and ionize other molecules, they increase the electron number density only linearly, not exponentially.

The dust current j_D is estimated simply, by the product of dust charge density $Q_{_L}$ and macroscopic motion $u_{_L}$, as:

$$j_D = Q_{_L} u_{_L}. \tag{78}$$

On the other hand the particle current j_p is determined by the Ohm's law:

$$j_p = \nu E_{\rm esf}, \tag{79}$$

where ν is the electric conductivity,

$$\nu = \frac{n_e\,l_{\rm mfp}\,e^2}{m_e\,v_e}. \tag{80}$$

E_{esf} is determined at the equilibrium of these two currents j_D and j_p:

$$j_D + j_p = 0. \tag{81}$$

By substituting (78), (79), and (80) into (81), we obtain

$$E_{\text{esf}} = -\frac{m_e\, v_e\, Q_L\, u_L}{n_e\, l_{\text{mfp}}\, e^2}. \tag{82}$$

Now that we know both E_{esf} and E_{crit}, the condition for electric discharge is

$$|E_{\text{esf}}| \geq E_{\text{crit}}. \tag{83}$$

By substituting (77) and (82) into (83) , we have the following form of the condition for electric discharge:

$$\left|\frac{Q_L}{Q_e}\right| \geq \frac{\Delta W_{\text{ion}}}{m_e v_e\, |u_L|}. \tag{84}$$

Within our parameter range of interest, the behavior of the left hand side of (84) as we increase $И$ is that it first keeps values much smaller than the right hand side and then it monotonically increases (c.f. Figure. 9). Thus there is a unique value of $И$ at which the equality for (84) holds. We define this value to be $И_{\text{crit}}$, the critical dust number density at which lightning takes place. Note that the condition doesn't depend on the detail of the electron stopping processes because we can eliminate l_{mfp} from the condition.

4.2. Numerical Simulations for Uniform-box Lightning Estimation

We now briefly explain the second numerical experiments, where we varied the set of input parameters, r_S, r_L, D_S, and D_L, and for each set of input parameters we calculated the critical dust number density $И_{\text{crit}}$ at which the lightning strikes. The numerical results strongly suggest that the parameter space (r_S, r_L, D_S, D_L) is subdivided into several regions, at each of which $И_{\text{crit}}$ is a simple analytic function of parameters (r_S, r_L, D_S, D_L).

The parameter ranges are

$$1.0 \times 10^{-4}\ \text{cm} < r_S < 1.0 \times 10^{2}\ \text{cm}, \tag{85}$$

$$1.0\ \text{cm} < r_L < 1.0 \times 10^{3}\ \text{cm}, \tag{86}$$

$$2.0 < D_S < 3.0, \tag{87}$$

$$2.0 < D_L < 3.0, \tag{88}$$

with additional constraints

$$r_S < r_L, \tag{89}$$

$$D_S < D_L, \tag{90}$$

$$1.0 < И < 1.0 \times 10^{6}. \tag{91}$$

Constraint (89) requires that the smaller dust is smaller than the larger dust. Constraint (90) comes from empirical fact that larger dust aggregates have experienced more compactification, and have higher fractal dimension [44, 51]. Constraint (91) is cutoff value of our computation.

The specific results of computations are rather bulky, so here we omit them. See [30] for the visualized results.

4.3. Analytic Formulae for Lightning Conditions

In this section we derive the analytic form of $\mathcal{N}_{\text{crit}}$ and lightning conditions. Numerical results obtained in §4.2. are of great help in deriving these analytic formulae. We show at the end of §4.3.4. that by our analytic formulae we can fit 364325 numerically-obtained points distributed among six decades with 21 per cent precision. Moreover, the formulae 'predicts' results of another simulation with 59 per cent precision, where charge exchange is 10^4 times inefficient. These agreements are good evidences for correctness of both numerical and analytical results.

We made plots like Figure 9 for many points within our parameter space, and found that $\mathcal{N} = \mathcal{N}_{\text{crit}}$ is met at the boundary of (c)charge-up phase and (d)dust phase in most cases, and sometimes in (d)dust phase or (c)charge-up phase. Therefore we derive analytic form of $\mathcal{N}$ corresponding to these three cases in §4.3.1. to §4.3.3., and combine them in §4.3.4..

4.3.1. Analytic Formulae for Charge-up Phase / Dust Phase Boundary

We first calculate $\mathcal{N}^{(cd)}$, the value of $\mathcal{N}$ corresponding to the (c)charge-up phase / (d)dust phase boundary.

The boundary satisfies $J_{L,S} = J_{i,S}$ (Table 2). We use the approximations in (c)charge-up phase to find the break point of the phase. Then

$$J_{L,S} \simeq 2\pi r_S r_L \Delta q_{A,C} n_S n_L \Delta v_{L,S} \tag{92}$$
$$J_{S,i} \simeq -Q_i n_S \pi r_S{}^2 \sigma_{S,i} v_i \tag{93}$$

because we can ignore the neutralization current $J^{(n)}_{L,S}$, approximate $\sigma_{L,S} = \pi r_L{}^2$, $\sigma_{S,i} = \pi r_S{}^2$ (geometric cross sections) and $\sigma_{L,i} = \pi r_L{}^2 |eQ_L| (r_L k_B T n_L)^{-1}$ ($qq'a^{-1} \gg k_B T$ limit of Coulomb cross section (32)).

In dust charge-up phase, both ions and electrons are mainly absorbed by smaller dust, so from (53) and (54) we have

$$Q_i = \frac{e\zeta n_g}{n_S \sigma_{S,i} v_i} \tag{94}$$
$$Q_e = \frac{-e\zeta n_g}{n_S \sigma_{S,e} v_e} \tag{95}$$

and the absorption cross sections are geometric : $\sigma_{S,i} = \sigma_{S,e} = \pi r_S{}^2$.

By substituting (94) into (93)

$$J_{S,i} = e\zeta n_g = J_{e,i} \tag{96}$$

cf. Figure 9(c) and in equation (70).

We have come to a simple result, that $\mathcal{N}_{\text{crit}}$ satisfies

$$J_{L,S} = J_{e,i}. \tag{97}$$

Substituting equations (92) , (96), together with $n_S = \mathcal{N}^{(cd)} n_S^{\bowtie}$ and $n_L = \mathcal{N}^{(cd)} n_L^{\bowtie}$ into equation (97) and solving for dust number density $\mathcal{N}$, we have

$$\mathcal{N}^{(cd)} = \left(\frac{\alpha^{(cd)}}{2\pi} \frac{e}{\Delta q_{A,C}} \frac{n_g}{n_S^{\bowtie} n_L^{\bowtie} r_S r_L} \frac{\zeta}{\Delta v_{L,S}} \right)^{\frac{1}{2}} \tag{98}$$

We have introduced a non-dimensional correction factor $\alpha^{(cd)}$, a constant that does not depend on r_S, r_L, D_S, D_L. We need this to compensate the error arising from using the formulae in (c)charge-up phase to find the break point of itself. The actual value for $\alpha^{(cd)}$ is in §4.3.4..

4.3.2. Analytic Formula for $И_{\rm crit}$ in Dust Phase

Next, we derive the analytic formula of the critical density $И^{(d)}_{\rm crit}$, where the condition for electric discharge (84) is met in (d)dust phase (c.f. §3.10.4., Figure 10(d)).

Imposing $J_{L,S} = 0$ in (45), and by approximating the charge neutrality (55) with $Q_S + Q_L = 0$, we have

$$Q_S = |Q_L| = \Delta q_{A,C} n_S \frac{2r_S}{r_L} \tag{99}$$

By approximating equation (54) with $J_{S,e} + J_{i,e} = 0$, we have

$$Q_e = -\frac{e n_g \zeta}{\pi n_S {r_S}^2 (1+\chi) v_e} \tag{100}$$

$$where\ 1+\chi = 1 + \frac{Q_S e}{n_S r_S k_B T}; \tag{101}$$

the factor $(1+\chi)$ comes from the Coulomb cross section (32).

Substituting Q_L and Q_e, the equality for the lightning condition (84) becomes

$$\frac{2\pi \Delta q_{A,C}}{e} \frac{{n_S}^2 {r_S}^3 (1+\chi)}{n_g r_L} \frac{v_e}{\zeta} = \frac{\Delta W_{\rm ion}}{m_e v_e u_L} \tag{102}$$

By substituting $n_S = И^{(d)}_{\rm crit} n^{\bowtie}_S$ and by solving for $И^{(d)}_{\rm crit}$, we have the following analytic formula for $И^{(d)}_{\rm crit}$:

$$И^{(d)}_{\rm crit} = \left(\frac{\alpha^{(d)}}{2\pi (1+\chi)} \frac{e}{\Delta q_{A,C}} \frac{n_g r_L}{n^{\bowtie 2}_S {r_S}^3} \frac{\zeta}{u_L} \frac{\Delta W_{\rm ion}}{k_B T} \right)^{\frac{1}{2}}. \tag{103}$$

We have introduced another non-dimensional correction constant $\alpha^{(d)}$ as we did in §4.3.1..

4.3.3. Analytic Formula for $И_{\rm crit}$ in Charge-up Phase

Finally, we derive the analytic formula of the critical density $И^{(c)}_{\rm crit}$, where the condition for electric discharge (84) is met in (c)charge-up phase (c.f. §3.10.3., Figure 10(c)).

By approximating equations (53) and (55) with $J_{S,i} = J_{i,e}$ and $Q_S + Q_i = 0$, we have

$$Q_i = -Q_S = \frac{e n_g \zeta}{\pi n_S {r_S}^2 v_i} \tag{104}$$

In equation (51), we can ignore $J_{L,e}$ and further ignoring the second term in equation (45), we have

$$\frac{2r_S}{r_L}\Delta q_{A,C} n_S n_L \Delta v_{L,S} \pi {r_L}^2 - Q_i n_L \sigma_{L,i} v_i = 0 \tag{105}$$

$$where\ \sigma_{L,i} = \frac{Q_L e}{n_L r_L l_B T} \pi {r_L}^2 \tag{106}$$

here we used the $qq'a^{-1} \gg k_B T$ limit of Coulomb cross section (32).

Solving this for Q_L, we have

$$|Q_L| = \frac{2\Delta q_{A,C}}{e^2} \frac{{n_S}^2 n_L {r_S}^3}{n_g} \frac{\Delta v_{L,S} k_B T}{\zeta} \tag{107}$$

And from equation (54) we have

$$Q_e = -\frac{e n_g \zeta}{\pi n_S {r_S}^2 v_e} \tag{108}$$

By substituting these Q_L and Q_e to the equality for the lightning condition (84), replacing $n_S = \text{И}^{(c)}_{\text{crit}} n^{\bowtie}_S$ and $n_L = \text{И}^{(c)}_{\text{crit}} n^{\bowtie}_L$, and by solving for $\text{И}^{(c)}_{\text{crit}}$, we obtain the following analytic formula for $\text{И}^{(c)}_{\text{crit}}$:

$$\text{И}^{(c)}_{\text{crit}} = \left(\frac{\alpha^{(c)}}{2\pi} \frac{e^3}{\Delta q_{A,C}} \frac{{n_g}^2}{n^{\bowtie 3}_S n^{\bowtie}_L {r_S}^5} \frac{\zeta^2}{\Delta v_{L,S} u_L} \frac{\Delta W_{\text{ion}}}{(k_B T)^2} \right)^{\frac{1}{4}} \tag{109}$$

We have introduced a third non-dimensional correction constant $\alpha^{(c)}$ as we did in previous sections.

4.3.4. The Combined Analytic Formula for И_{crit}

The critical number density И_{crit} is either of $\text{И}^{(c)}_{\text{crit}}$, $\text{И}^{(cd)}$, $\text{И}^{(d)}_{\text{crit}}$. To choose the correct one, we have to consider the phase boundary conditions (c.f. Table 2). Instead, we propose the following convenient scheme to choose the correct one:

$$\begin{aligned} \text{И}_{\text{crit}} &= \text{И}^{(d)}_{\text{crit}} \quad if\ \text{И}^{(d)}_{\text{crit}} > \text{И}^{(cd)}, \\ &= \text{И}^{(cd)} \quad if\ \text{И}^{(c)}_{\text{crit}} > \text{И}^{(cd)} > \text{И}^{(d)}_{\text{crit}}, \\ &= \text{И}^{(c)}_{\text{crit}} \quad otherwise. \end{aligned} \tag{110}$$

This scheme is based on the intuition that the (cd)phase boundary is included in both (c)charge-up phase and (d)dust phase. We can argue that if $\text{И}^{(d)}_{\text{crit}} > \text{И}^{(cd)}$, the number density $\text{И}^{(cd)}$ is not large enough to cause lightning, and that if $\text{И}^{(c)}_{\text{crit}} < \text{И}^{(cd)}$, the number density $\text{И}^{(c)}_{\text{crit}}$ is already large enough to cause lightning. Now, without the correction, e.g. with $\alpha^{(c)} = \alpha^{(cd)} = \alpha^{(d)} = 1$, the analytic values for И_{crit} differs from the numerical values $\text{И}^{(num)}_{\text{crit}}$ calculated in §4.2., because of approximations used. We set the values for

$\alpha^{(c)}, \alpha^{(cd)}, \alpha^{(d)}$ by the condition that the following squared-error integral over the entire parameter ranges (85-91) is minimum:

$$\iiiint dr_S\, dr_L\, dD_S\, dD_L \left(\log_{10} И_{\rm crit} - \log_{10} И_{\rm crit}^{(num)}\right)^2 \tag{111}$$

This gives $\alpha^{(c)} = 9.4 \times 10^{-1}, \alpha^{(cd)} = 3.3 \times 10^{-1}, \alpha^{(d)} = 8.5 \times 10^{-1}$. Taking these corrections into account, the values for $И^{(c)}, И^{(cd)}, И^{(d)}$ are as follows:

$$\begin{aligned}
И_{\rm crit}^{(c)} &= 1.1 \times 10^3 \left(\frac{\Delta q_{A,C}}{6.2 \times 10^2 \text{ e}}\right)^{-\frac{1}{4}} \left(\frac{n_g}{4.7 \times 10^{13} \text{ cm}^{-3}}\right)^{\frac{1}{2}} \\
&\left(\frac{n_S^{\bowtie}}{8.8 \times 10^{-1} \text{ cm}^{-3}}\right)^{-\frac{3}{4}} \left(\frac{n_L^{\bowtie}}{4.0 \times 10^{-14} \text{ cm}^{-3}}\right)^{-\frac{1}{4}} \\
&\left(\frac{r_S}{1.0 \times 10^{-4} \text{ cm}}\right)^{-\frac{5}{4}} \\
&\left(\frac{\zeta}{1.0 \times 10^{-18} \text{ sec}^{-1}}\right)^{\frac{1}{2}} \\
&\left(\frac{\Delta v_{L,S}}{3.4 \times 10^3 \text{ cm sec}^{-1}}\right)^{-\frac{1}{4}} \left(\frac{u_L}{3.4 \times 10^3 \text{ cm sec}^{-1}}\right)^{-\frac{1}{4}} \\
&\left(\frac{\Delta W_{\rm ion}}{15.4 \text{ eV}}\right)^{\frac{1}{2}} \left(\frac{T}{1.7 \times 10^2 \text{ K}}\right)^{-\frac{1}{2}}, \qquad (112) \\
И^{(cd)} &= 3.3 \times 10^2 \left(\frac{\Delta q_{A,C}}{6.2 \times 10^2 \text{ e}}\right)^{-\frac{1}{2}} \left(\frac{n_g}{4.7 \times 10^{13} \text{ cm}^{-3}}\right)^{\frac{1}{2}} \\
&\left(\frac{n_S^{\bowtie}}{8.8 \times 10^{-1} \text{ cm}^{-3}}\right)^{-\frac{1}{2}} \left(\frac{n_L^{\bowtie}}{4.0 \times 10^{-14} \text{ cm}^{-3}}\right)^{-\frac{1}{2}} \\
&\left(\frac{r_S}{1.0 \times 10^{-4} \text{ cm}}\right)^{-\frac{1}{2}} \left(\frac{r_L}{1.0 \text{ cm}}\right)^{-\frac{1}{2}} \\
&\left(\frac{\zeta}{1.0 \times 10^{-18} \text{ sec}^{-1}}\right)^{\frac{1}{2}} \\
&\left(\frac{\Delta v_{L,S}}{3.4 \times 10^3 \text{ cm sec}^{-1}}\right)^{-\frac{1}{2}}, \qquad (113)
\end{aligned}$$

$$\begin{aligned}
И^{(d)}_{\rm crit} &= 5.9 \times 10^1 \left(\frac{\Delta q_{A,C}}{6.2 \times 10^2 \text{ e}}\right)^{-\frac{1}{2}} \left(\frac{n_g}{4.7 \times 10^{13} \text{ cm}^{-3}}\right)^{\frac{1}{2}} \\
&\left(\frac{n_S^{\bowtie}}{8.8 \times 10^{-1} \text{ cm}^{-3}}\right)^{-1} \\
&\left(\frac{r_S}{1.0 \times 10^{-4} \text{ cm}}\right)^{-\frac{3}{2}} \left(\frac{r_L}{1.0 \text{ cm}}\right)^{\frac{1}{2}} \\
&\left(\frac{\zeta}{1.0 \times 10^{-18} \text{ sec}^{-1}}\right)^{\frac{1}{2}} \left(\frac{u_L}{3.4 \times 10^3 \text{ cm sec}^{-1}}\right)^{-\frac{1}{2}} \\
&\left(\frac{\Delta W_{\rm ion}}{15.4 \text{ eV}}\right)^{\frac{1}{2}} \left(\frac{T}{1.7 \times 10^2 \text{ K}}\right)^{-\frac{1}{2}} . \qquad (114)
\end{aligned}$$

Note that $\Delta q_{A,C}$, $n_S^{\bowtie}$, and $n_L^{\bowtie}$ also depends on dust parameters: r_S, r_L, D_S, and D_L. Using equation (39) and the $И_{ch}, \sigma_{ch}$ introduced in §3.7.1.,

$$\begin{aligned}
\Delta q_{A,C} &= И_{ch} \sigma_{ch} S_{\rm kiss} \qquad (115) \\
&= 6.2 \times 10^2 \text{ e} \cdot \frac{И_{ch}}{0.1} \frac{\sigma_{ch}}{6.2 \times 10^9 \text{ e cm}^{-2}} \\
&\frac{\min\left({r_S}^{3/2} {r_L}^{1/2}, {r_m}^{2-D_L} {r_S}^{5/2} {r_L}^{D_L - 5/2}\right)}{1.0 \times 10^{-6} \text{ cm}^2}. \qquad (116)
\end{aligned}$$

Using equations (6-12) and equation (30),

$$\begin{aligned}
n_S^{\bowtie} &= 4.0 \times 10^2 \left(\frac{r_S}{r_m}\right)^{-D_S} \text{ cm}^{-3} \\
&\left(\frac{r}{2.7 \text{ AU}}\right)^{-11/4} \left(\frac{m_m}{3.9 \times 10^{-15} \text{ g}}\right)^{-1} \qquad (117) \\
n_L^{\bowtie} &= 4.0 \times 10^1 \left(\frac{r_L}{r_m}\right)^{-D_L} \text{ cm}^{-3} \\
&\left(\frac{r}{2.7 \text{ AU}}\right)^{-11/4} \left(\frac{m_m}{3.9 \times 10^{-15} \text{ g}}\right)^{-1} , \qquad (118)
\end{aligned}$$

and the monomer radius

$$r_m = 1.0 \times 10^{-5} \text{ cm}. \qquad (119)$$

These analytic solutions (112-114) combined with the condition (110) are in excellent agreement with the numerical results. Quantitatively, the root-mean-square error is

$$\sqrt{\frac{\iiiint dr_S \, dr_L \, dD_S \, dD_L \left(\log_{10} И_{\rm crit} - \log_{10} И^{(num)}_{\rm crit}\right)^2}{\iiiint dr_S \, dr_L \, dD_S \, dD_L}} = 9.2 \times 10^{-2}. \qquad (120)$$

Moreover, using the values of $\alpha^{(c)}$, $\alpha^{(cd)}$, $\alpha^{(d)}$ obtained only from the 'normal' run , we can fit the results of the 'pessimistic' simulations [1] by a root-mean-square error of 2.6×10^{-1}. We also perform the simulations with smaller values of relative velocity $\Delta v_{L,S}$ and fit the results. The root-mean-square errors were $5.6 \times 10^{-2}, 6.4 \times 10^{-2}, 1.1 \times 10^{-1}$, for $\Delta v_{L,S} = u_L = 3.4 \times 10^2$ cm s^{-1}, 3.4×10^1 cm s^{-1}, 3.4 cm s^{-1}, respectively. These fits prove the predictability of our analytic formulae (110) and (112-114).

4.4. Conclusion of the Section — Uniform-box Model for Protoplanetary Disk Lightning

We have calculated the dust number density $И_{\rm crit}$ at which lightning strikes, as function of dust radius r_S, r_L and fractal dimension D_S, D_L numerically. Using the numerical results we have derived the analytical formulae for $И_{\rm crit}$: equations (110), (112-114). Because the generated electrostatic field $E_{\max}(\eta)$ grows more rapidly than estimate by Gibbard et al.[15] in (c)charge-up phase and (d)dust phase, lightning in protoplanetary discs are possible with smaller dust number densities.

5. Resistive Magnetohydrodynamic Simulation — Towards the Morphology of Disk Lightning

5.1. The Two-dimensional Model for Disk Lightning

Now that we know the one-zone behavior of the dust plasma, the natural question arises next is "how does the lightning look like?" To answer this question, to provide quantitative predictions such as event ratio, energy spectrum and morphology of the lightning, we must combine the dust-plasma model with fluid models that depict the protoplanetary disk. We are now on the way to do this, by combining the analytical solutions (eq 110, 112-114) with computational magnetohydrodynamic simulations. Our first step, a two-dimensional disk lightning model is described in this section.

Figure 12 illustrates the two-dimensional model of the $\phi - z$ slice of the disk. Two fluids are treated separately: one is the gas component coupled with small dust, the other is large dust. Gravity is towards the equatorial plane. Initial velocity shear is given to cause Kelvin-Helmholtz instability. Thus shear velocity is given, rather than self-maintained, for simplicity.

Magnetohydrodynamics (MHD) is the governing equations for ionized gas. Especially if you want to study lightning, you need electrostatic field. To have nonzero electric field in the comoving frame of the fluid, you must go to resistive MHD:

[1] the charge exchange efficiency $И_{ch}$ is set to be 10^4 times smaller value

$$\frac{\partial \rho}{\partial t} + \nabla \cdot (\rho \mathbf{u}) = 0, \tag{121}$$

$$\frac{\partial \rho \mathbf{u}}{\partial t} + \nabla \cdot (\rho \mathbf{u} \otimes \mathbf{u} - \mathbf{B} \otimes \mathbf{B}) + \nabla P = 0, \tag{122}$$

$$\frac{\partial \mathbf{B}}{\partial t} + \nabla \times \mathbf{E} = 0, \tag{123}$$

$$\frac{\partial e}{\partial t} + \nabla \cdot ((e + P)\, \mathbf{u} + \mathbf{E} \times \mathbf{B}) = 0. \tag{124}$$

Here, independent variables are ρ: density, $\rho\mathbf{u}$: momentum density, $\mathbf{B}$:magnetic field, and e: energy density. The degree of freedom is 7 due to a constraint from Maxwell's equation

$$\nabla \cdot \mathbf{B} = 0. \tag{125}$$

Other variables, such as $\mathbf{j}$: electric current, $\mathbf{E}$: electric field, P: gas pressure etc. are dependent variable in the framework of MHD, as follows:

$$\mathbf{j} = \nabla \times \mathbf{B}, \tag{126}$$

$$\mathbf{E} = -\mathbf{u} \times \mathbf{B} + \eta \mathbf{j}, \tag{127}$$

$$P = (\gamma - 1)\left(e - \frac{1}{2}\rho u^2 - \frac{1}{2}B^2\right), \tag{128}$$

where γ is the ratio of specific heats and η is the magnetic resistivity. We used multidimensional mesh-based MHD code "ATHENA" by Stone et al. [43], and modified the code for our purpose, by adding magnetic resistivity, advected dust, and discharge model.

Advected scalar fields $S[i]$ can be easily added to this model:

$$\frac{\partial S[i]}{\partial t} + \nabla \cdot (S[i]\mathbf{u}) = 0. \tag{129}$$

We add two scalar fields: $S[0]$ is the number density of the larger dust, and $S[1]$ is the non-equilibrium ionization degree. We need $S[0]$ to generate the electric field, and $S[1]$ to calculate the magnetic resistivity η, and to make lightning branch. The specific equations for the scalar fields are:

$$\frac{\partial S[0]}{\partial t} + \nabla \cdot (S[0]\,(\mathbf{u} + \mathbf{u}_g(r))) = 0, \tag{130}$$

$$\frac{\partial S[1]}{\partial t} + \nabla \cdot (S[1]\mathbf{u} - D_1 \nabla S[1]) + \frac{S[1] - S_{eq}[1]}{\tau_1} = \delta_{discharge}. \tag{131}$$

Large dust advects while falling at the terminal velocity $\mathbf{u}_g(r)$. Ionization suddenly rise at lightning path (discharge model; mentioned next) then it advects, diffuses with diffusion coefficient D_1, and rapidly decays with timescale τ_1 to equilibrium value $S_{eq}[1]$.

Finally, we describe our numerical discharge model (c.f. Figure 11.), based on the lightning model by Solomon et al. [40].

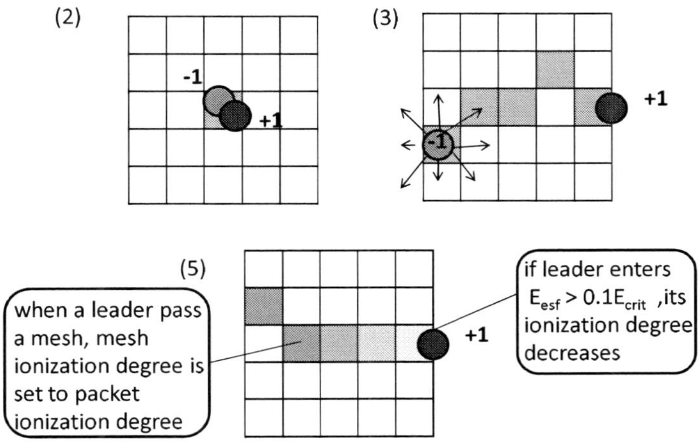

Figure 11. The model for lightning leader propagation.

1. Suppose the dust plasma is in (c)charge-up phase: Electric field is dominated by ηj term.

2. Introduce a pair of lightning leader packet ($q = \pm 1$) at every mesh when it satisfies $E_{esf} > E_{crit}$. q is the sign of the electric charge carried by the leader.

3. Each lightning leader random-walks as a chess king. They obey electric field and prefers the highly ionized mesh. The probability $P(\mathbf{n})$ that direction $\mathbf{n}$ is chosen satisfies

$$P(\mathbf{n}) \propto \mathrm{expone}\left(qS[1]^{0.5}\frac{\mathbf{n} \cdot \mathbf{E}_{\mathrm{esf}}}{|E_{\mathrm{esf}}|}\right), \tag{132}$$

   where expone is a function such that $\mathrm{expone}(x) = x$ for $x > 0$ and $\mathrm{expone}(x) = \exp(x)$ otherwise.

4. Initially, leader packet has ionization degree $S[1] = 1.0$

5. At each mesh the lightning leader passes, mesh ionization degree is set to the leader ionization degree. Leader ionization is updated as

$$S[1] \leftarrow S[1] \times \min\left(1.0, \frac{|E_{\mathrm{esf}}|}{0.1E_{\mathrm{crit}}}\right). \tag{133}$$

   Lightning leader packet can survive as long as $E_{\mathrm{esf}} > 0.1E_{\mathrm{crit}}$ (not $E_{\mathrm{esf}} > E_{\mathrm{crit}}$). This is a result of the avalanche discharge model; once generated, lightning leader can maintain the ionizing process even if the electric field is substantially lower than the critical value.

6. If the leader ionization degree is smaller than the mesh ionization degree, remove the leader packet.

7. This happens in a instant, compared to hydrodynamic timescales.

5.2. The Simulation Result

The time evolution for this model is depicted by frames in Figure 13. As time goes by; (1,2): The Kelvin-Helmholtz instability initiates and develops; (3): The first lightning takes place; (4,5,6): As the large dust sediments, its number density is increased and the lightning events become gradually dense and frequent; (7,8): At the late phase, a lot of lightning takes place in parallel; (9): At the final phase, the dust layer become thin and ionization degree is maintained high, so the lightning become few and powerless.

Although many things are left to do in this simulations to reach reality — the turbulence is initial condition driven instead of self-maintained, set of parameters are deliberately chosen values[2] for the favor of lightning instead of realistic values — the computer program used here demonstrates the possibility of, and can actually used in, future studies of protoplanetary disk lightning, for its morphology, event rate and energy spectrum.

6. Possible Observations

We have assumed a uniform region in a protoplanetary disk and expressed the dust charge Q_{I}, and the generated electric field E_{esf}, as functions of dust number density $И$, dust radius r_S, r_L and fractal dimension D_S, D_L. In turn, we calculated the critical number density $И_{\mathrm{crit}}$ at which lightning strikes as functions of dust parameters r_S, r_L, D_S and D_L. Based on these knowledge, we have put a hand to studies of the morphology of the disk lightning. We now know that dust-dust collisional charge separation do contribute to protoplanetary disk physics by creating new phases of the dust-plasma, and that disk lightning is plausible. Let us see now, what we disk lightning can say to the astronomical observations.

6.1. Energetics and Direct Observations

We estimate the total energy of a lightning event in a protoplanetary disc at $r = 2.7$ AU. For MMSN, the number density of the gas is 4.7×10^{13} cm^{-3} in the region. The typical electron mean free path at this site is $l_{mfp} \simeq 1.2 \times 10^2$ cm. By equation (77) we know the critical electric field $E_{\mathrm{crit}} \simeq 4.3 \times 10^{-4}$ G. The sphere with radius of the disc scale-height $h \simeq 2.4 \times 10^{12}$ cm contains the electric energy $W \equiv {E_{\mathrm{crit}}}^2/8\pi \times 4\pi h^3/3 \simeq 4.3 \times 10^{29}$ erg. When the lightning strikes, the energy is concentrated into lightning bolt of radius w and length h, where w is related to l_{mfp} by $w \simeq 5000\, l_{mfp} \simeq 6.0 \times 10^5$ cm [35]. If all the energy is used to heat the gas within the lightning bolt, the gas can be heated to 1.6×10^7 K.

The ultimate energy source for this electric discharge event is the gravitational energy of the accreting matter. In our model the mass accretion ratio of uncondensed larger dust is $\dot{M} = 2\pi r \Sigma_L^{\bowtie} u_L \simeq 3.3 \times 10^{17}$ g sec^{-1}. The gravitational energy released within condensation region h is $L \equiv G M_{\odot} \dot{M} h r^{-2} \simeq 6.6 \times 10^{28}$ erg sec^{-1}. For the largest energy event $W = 4.3 \times 10^{29}$ erg, The upper limit of the event rate is 1.5×10^{-1} sec^{-1}.

[2]Although, they are in the range plausible for protoplanetary disks.

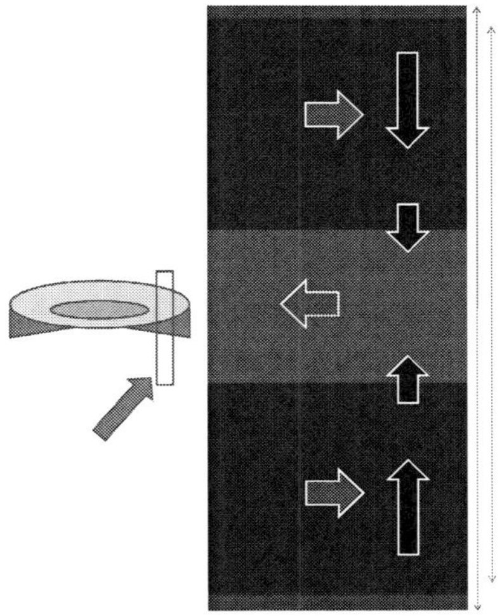

Figure 12. The two dimensional model of the cross section of the the protoplanetary disk.

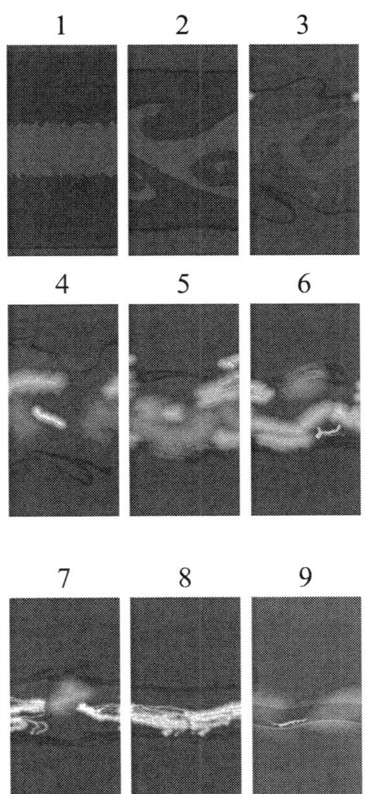

Figure 13. The two-dimensional simulation of the disk lightning.

6.1.1. Astronomically Low Frequency (ALF) Waves

The change density evolution, electromagnetic pulse, and electromagnetic waves accompanying lightning in terrestrial thunderclouds are observed [e.g. 22, 26]. The typical wavelength of the electromagnetic waves are similar to the scale height of the thundercloud. These are called extremely low frequency waves. The electromagnetic waves from lightning can be basically modeled as solutions of Maxwell equations, including lightning current as a source term [e.g. 36]. When we apply these models to the protoplanetary discs, the electromagnetic wave spectrum is extend between the event duration and light crossing time of the system, or $9.6 \times 10^{-5} \sim 1.2 \times 10^{-2}$ Hz. This frequency range is at least two orders of magnitude lower than any frequencies with established observational methods. We propose the term 'Astronomically Low Frequency (ALF) waves' for such waves, the successor to the frequency list 'very low frequency (*VLF*),' 'ultra low frequency(*ULF*),' 'super low frequency (*SLF*),' and 'extremely low frequency (*ELF*).' The frequency is so low that we will need an astronomical budget to build an astronomically large detector to receive it, considering its wavelength of order of an astronomical unit.

6.1.2. Infrared (IR) Observations

The energy of the lightning contributes to the local heating of the protoplanetary discs, which might be resolved by advanced telescopes such as Atacama Large Millimeter Array (*ALMA*). The most possible observational evidence is excess of heating near the snowline. To distinguish the cause of the heating with other heating model candidates, the variability or correlation function of the heating might be useful. This is because lightning propagates at the speed of ionized electrons, which is much faster than the speed of sound.

6.1.3. Ultraviolet (UV) Observations

The ionization electrons of the lightning excite various electron levels in gas molecules and dust. There is possibility of observing fluorescence photons from such excited molecules. Although the disc gas is generally expected to be thick for ultraviolet photons, there are categories of lightning that extends toward thin regions of the gas, known as sprites and elves [e.g. 53]. The sprites and elves are phenomena similar to lightning observed in the mesosphere of the earth, possibly caused by electric fields induced by the thunderclouds. Fluorescence lines from such regions can be observed by future ultraviolet missions like *THEIA* [41]. Also, some observational results on protostellar and protoplanetary systems today have difficulties in explaining either lack or excess of *UV* [e.g. 32, 6, 34, 19]. If excess of *UV* photons is observed compared to the model, it might be from the sprite discharges and elves from the surface of the protoplanetary discs; on the other hand if the chemical composition model require more *UV* photons than is observed, lightning hidden in the disc mid-plane might be providing them.

6.1.4. High Energy Gamma Rays

Detection of burst-like gamma-ray is reported from terrestrial thunder clouds. The burst precedes a cloud-to-ground lightning, lasts for $\sim$ 40 seconds, extends to 10 MeV. The spectrum can be interpreted as consisting of bremsstrahlung photons from relativistic electrons [48, 12]. These relativistic electrons are secondary electrons generated by cosmic rays, and accelerated by the electric fields through process known as avalanche amplification [37]. If a charged particle is accelerated by the protoplanetary thundercloud field, through similar process, its kinetic energy reaches $eEh \simeq 3.1 \times 10^{11}$ eV.

6.2. Implication on Chondrule Formations

In §3.1., we have introduced chondrule and its formation, and have addressed three problems against explaining the chondrule formation by lightning.

This work can provide answer for the first and second problem. For the energetics problem, the larger dust and the gas (containing smaller dust that are coupled to the gas) is now 'harnessed' by electric field. Outgoing gas is not free in carrying the gravitational energy away; instead the gas converts its gravitational energy into electric field energy, fully contributing to lightning. For the neutralization problem, we have shown in this work that with reasonably high dust number density η, the dust-dust charge separation can dominate over the plasma neutralization, and the electrostatic field can grow up to critical value.

For the third problem, we point out that in Güttler[16]'s experiment, either the electron mean free path is many orders of magnitude shorter, or the electron kinetic energy is much larger compared to the protoplanetary-disc environment. They used air at pressures between 10 and 10^5 Pa. Air consists of 78 per cent nitrogen, 21 per cent oxygen, and 1 per cent argon. Their molecular van der Waals radii are 1.6×10^{-8} cm, 1.5×10^{-8} cm, and 1.9×10^{-8} cm, respectively [4].

Therefore, the electron mean free path and the electron kinetic energy, $W_e = e\,E\,l_{\rm mfp}$, was $l_{\rm mfp} \sim 4.8 \times 10^{-1}$ cm, $W_e = 1.6 \times 10^4$ eV for 10 Pa case, and $l_{\rm mfp} \sim 4.8 \times 10^{-5}$ cm , $W_e = 1.6$ eV for 10^5 Pa case, respectively. On the other hand in protoplanetary discs, typical mean free path and electron kinetic energy are $l_{\rm mfp} = 1.2 \times 10^2$ cm, $W_e = 15.4$ eV.

It might be possible that protoplanetary-disc lightning is effective in melting dust aggregates, although experimental lightning is ineffective in heating and led to disruption of the dust, due to shorter mean free path or higher energy electron. The minimum size of the structures that electron can form is of order of its mean free path. If the electron mean free path is much shorter than the dust aggregates, as in 10^5 Pa case, the electron current may concentrate on the most conductive part of the dust aggregate, leading to partial heating and explosion of the dust. On the other hand if the electron is much more energetic, as in 10 Pa case, it may react differently on dust monomers.

To reproduce the mean free path and electron energy simultaneously, one must reproduce the electric field strength $E = 4.3 \times 10^{-4}$ G of protoplanetary discs; while the electric field used in the experiment $E = 1.1 \times 10^2$ G was much stronger. This much stronger electric field itself, might be the cause of dust aggregate dissociation, due to much stronger electric force exerted on electron-absorbed dust monomers. Also the discharge time-scale in the experiment was much smaller than that in the protoplanetary discs, which might have led to the catastrophic results.

We think that the effect of lighting on dust aggregate in protoplanetary-disc environment is yet to be confirmed in future experiments and simulations.

6.3. Limitations to the Observations

Chondrule heating by shockwave model, however, remains plausible. One can say that if lightning is restricted to certain region e.g. the snow line of a protoplanetary disk, the chondrule formation model by lightning cannot explain the ubiquity of the chondrule. [3]

The strength of the ALF wave signal from an 100 pc distant disk, can be estimated as 6.6×10^{28} erg s$\times5\%\times1/4\pi\times(100\,\mathrm{pc})^{-2}\times(10^{-2}\,\mathrm{Hz})^{-1} = 2.9\times10^{10}$ Jy for 10^{-2} Hz band. Here 5% is a crude estimate for a conversion efficiency. The signal intensity itself is not too bad compared to other sources. However, not to mention the observational difficulties caused by the too long wavelength, plasma structure such as the heliopause might shield such long waves and we may have to resort to trans-stellar measures just to put upper limits of the signal [24]. The heating provided from the lightning is orders of magnitude smaller than the thermal heating, and UV signals are known to be absorbed by the interstellar dust. In the end, most plausible means of observing the disk lightning might be to "see" them, in optical band. Moreover, it can happen that "lightning-luminous" era is, for example, 100 years, within the total 1000'000 years of the history of a protoplanetary disk. In that case, the possibility that one of the disk found in our neighbor is in the lightning era is very low.

7. Conclusion

We have discussed several new aspects of planet formation theory, when dust charge is taken into account. The dust charge brought by plasma absorption may drastically alter the planet formation scenario at its beginning; the planet formation is not in-place nor homogeneous. Instead, planet seeds migrate all across the protoplanetary disk to grow up. Furthermore, the charge brought in by dust-dust collision has an important effect. It does have non-negligible effect on the charge state of the dust and the plasma, and at certain environments, it may cause lightning in protoplanetary disks. We are now preparing to study the disk lightning quantitatively. If such astronomical lightning does take place and one day is observed, we will read-out unique information on the mysterious protoplanetary disks. Even if, in the end, such lightning does not exist, we will learn a lot in the theoretical and observational pursuit.

References

[1] S. A. Balbus and J. F. Hawley. A powerful local shear instability in weakly magnetized disks. I - Linear analysis. II - Nonlinear evolution. *Astrophysical Journal*, 376:214–233, July 1991.

[2] Steven A. Balbus and John F. Hawley. Instability, turbulence, and enhanced transport in accretion disks. *Rev. Mod. Phys.*, 70(1):1–53, Jan 1998.

[3] J. Blum and G. Wurm. Experiments on Sticking, Restructuring, and Fragmentation of Preplanetary Dust Aggregates. *Icarus*, 143:138–146, January 2000.

[3] This criticism also apply to shock model, since strong shocks may only accompany gas-giant planets. Again, we must pay attention on the dynamism of the protoplanetary disk, which might have spread the chondrules.

[4] A. Bondi. van der Waals Volumes and Radii. *J. Phys. Chem.*, 68:441–451, March 1964.

[5] F. Brauer, C. P. Dullemond, and T. Henning. Coagulation, fragmentation and radial motion of solid particles in protoplanetary disks. *Astronomy & Astrophysics*, 480:859–877, March 2008.

[6] E. Chapillon, S. Guilloteau, A. Dutrey, and V. Piétu. Disks around CQ Tauri and MWC 758: dense PDR or gas dispersal? *Astronomy & Astrophysics*, 488:565–578, September 2008.

[7] H. Christian, C. R. Holmes, J. W. Bullock, W. Gaskell, A. J. Illingworth, and J. Latham. Airborne and ground-based studies of thunderstorms in the vicinity of Langmuir Laboratory. *The Quarterly Journal of the Royal Meteorological Society*, 106:159–174, December 1980.

[8] J. N. Cuzzi and K. J. Zahnle. Material Enhancement in Protoplanetary Nebulae by Particle Drift through Evaporation Fronts. *Astrophysical Journal*, 614:490–496, October 2004.

[9] J.G. Dash, B.L. Mason, and J.S. Wettlaufer. Theory of charge and mass transfer in ice-ice collisions. *Journal of Geophysical Research*, 106:20395–20402, January 2001.

[10] S. J. Desch and J. N. Cuzzi. The Generation of Lightning in the Solar Nebula. *Icarus*, 143:87–105, January 2000.

[11] W. W. Duley and D. A. Williams. *Interstellar chemistry*. 1984.

[12] T. Enoto, H. Tsuchiya, S. Yamada, and et al. Detection of gamma-rays from winter thunderclouds along the coast of Japan Sea. In *International Cosmic Ray Conference*, volume 1 of *International Cosmic Ray Conference*, pages 745–748, 2008.

[13] C. F. Gammie. Layered Accretion in T Tauri Disks. *Astrophysical Journal*, 457:355–+, January 1996.

[14] W. Gaskell, A.J. Illingworth, J. Latham, and C.B. Moore. Airborne studies of electric fields and the charge and size of precipitation elements in thunderstorms. *The Quarterly Journal of the Royal Meteorological Society*, 104:447–460, December 1978.

[15] S. G. Gibbard, E. H. Levy, and G. E. Morfill. On the Possibility of Lightning in the Protosolar Nebula. *Icarus*, 130:517–533, December 1997.

[16] C. Güttler, T. Poppe, J. T. Wasson, and J. Blum. Exposing metal and silicate charges to electrical discharges: Did chondrules form by nebular lightning? *Icarus*, 195:504–510, May 2008.

[17] C. Hayashi. Structure of the Solar Nebula, Growth and Decay of Magnetic Fields and Effects of Magnetic and Turbulent Viscosities on the Nebula. *Progress of Theoretical Physics Supplement*, 70:35–53, 1981.

[18] C. Hayashi, K. Nakazawa, and Y. Nakagawa. Formation of the solar system. In D. C. Black & M. S. Matthews, editor, *Protostars and Planets II*, pages 1100–1153, 1985.

[19] G. J. Herczeg and L. A. Hillenbrand. UV Excess Measures of Accretion onto Young Very Low Mass Stars and Brown Dwarfs. *Astrophysical Journal*, 681:594–625, July 2008.

[20] P. Kalas, J. R. Graham, E. Chiang, M. P. Fitzgerald, M. Clampin, E. S. Kite, K. Stapelfeldt, C. Marois, and J. Krist. Optical Images of an Exosolar Planet 25 Light-Years from Earth. *Science*, 322:1345–, November 2008.

[21] E. Kokubo and S. Ida. Oligarchic Growth of Protoplanets. *Icarus*, 131:171–178, January 1998.

[22] W.J. Koshak and E.P. Krider. Analysis of lightning field changes during active florida thunderstorms. *Journal of Geophisical Research*, 94:1165–1186, 1989.

[23] Konstantin N. Kudin and Roberto Car. Why are water-hydrophobic interfaces charged? *Journal of the American Chemical Society*, 130(12):3915–3919, 2008.

[24] B. C. Lacki. The End of the Rainbow: What Can We Say About the Extragalactic Sub-Megahertz Radio Sky? *ArXiv e-prints*, April 2010.

[25] A.-M. Lagrange, D. Gratadour, G. Chauvin, T. Fusco, D. Ehrenreich, D. Mouillet, G. Rousset, D. Rouan, F. Allard, É. Gendron, J. Charton, L. Mugnier, P. Rabou, J. Montri, and F. Lacombe. A probable giant planet imaged in the β Pictoris disk. VLT/NaCo deep L'-band imaging. *Astronomy & Astrophysics*, 493:L21–L25, January 2009.

[26] Y. T. Lin, M. A. Uman, J. A. Tiller, R. D. Brantley, W. H. Beasley, E. P. Krider, and C. D. Weidman. Characterization of lightning return stroke electric and magnetic fields from simultaneous two-station measurements. *Journal of Geophisical Research*, 84:63076314, 1979.

[27] C. Marois, B. Macintosh, T. Barman, B. Zuckerman, I. Song, J. Patience, D. Lafrenière, and R. Doyon. Direct Imaging of Multiple Planets Orbiting the Star HR 8799. *Science*, 322:1348–, November 2008.

[28] B.L. Mason and J.G. Dash. Charge and mass transfer in ice-ice collisions: Experimental observations of a mechanism in thunderstorm electrification. *Journal of Geophysical Research*, 105:10185–10192, 2000.

[29] H. Miura, T. Nakamoto, and M. Doi. Origin of three-dimensional shapes of chondrules. I. Hydrodynamics simulations of rotating droplet exposed to high-velocity rarefied gas flow. *Icarus*, 197:269–281, September 2008.

[30] T. Muranushi. Dust-dust collisional charging and lightning in protoplanetary discs. *Monthly Notices of the Royal Astronomical Society*, 401:2641–2664, February 2010.

[31] T. Muranushi and T. Tomiyasu. Proto-planetary Nebulae Struck by Lightning? A Charge Separation Study of Ice Dust Grains And its Effect on Dust Growth. In T. Usuda, M. Tamura, & M. Ishii, editor, *American Institute of Physics Conference Series*, volume 1158 of *American Institute of Physics Conference Series*, pages 145–146, August 2009.

[32] H. Nomura and T. J. Millar. Molecular hydrogen emission from protoplanetary disks. *Astronomy & Astrophysics*, 438:923–938, August 2005.

[33] S. Okuzumi. Electric Charging of Dust Aggregates and its Effect on Dust Coagulation in Protoplanetary Disks. *Astrophysical Journal*, 698:1122–1135, June 2009.

[34] M. R. Pérez, B. McCollum, M. E. van den Ancker, and M. D. Joner. The enigmatic young object: Walker 90/V590 Monocerotis. *Astronomy & Astrophysics*, 486:533–544, August 2008.

[35] W. Pilipp, T. W. Hartquist, and G. E. Morfill. Large electric fields in acoustic waves and the stimulation of lightning discharges. *Astrophysical Journal*, 387:364–371, March 1992.

[36] V. A. Rakov and M. A. Uman. Review and evaluation of lightning return stroke models including some aspects of their application. *IEEE Transactions on Electromagnetic Compatibility*, 40:403–426, 1998.

[37] R. Roussel-Dupré and A. V. Gurevich. On runaway breakdown and upward propagating discharges. *Journal of Geophysical Research*, 101:2297–2312, February 1996.

[38] D. M. Shaw, M. D. Higgins, R. W. Hinton, M. G. Truscott, and T. A. Middleton. Boron contamination in polished thin sections of meteorites. *American Mineralogist*, 73:894–900, 1988.

[39] S. Sirono. Effects by sintering on the energy dissipation efficiency in collisions of grain aggregates. *Astronomy & Astrophysics*, 347:720–723, July 1999.

[40] Robert Solomon, Claudia Adamo, and Marcia Baker. A lightning initiation mechanism: application to a thunderstorm electrification model. *Comptes Rendus Physique*, 3(10):1325 – 1333, 2002.

[41] D. N. Spergel, J. Kasdin, and R. et al. Belikov. THEIA: Telescope for Habitable Exoplanets and Interstellar/Intergalactic Astronomy. In *Bulletin of the American Astronomical Society*, volume 41 of *Bulletin of the American Astronomical Society*, pages 362–+, January 2009.

[42] L. J. Spitzer. The Dynamics of the Interstellar Medium. I. Local Equilibrium. *Astrophysical Journal*, 93:369–+, May 1941.

[43] J. M. Stone, T. A. Gardiner, P. Teuben, J. F. Hawley, and J. B. Simon. Athena: A New Code for Astrophysical MHD. *Astrophysical Journals*, 178:137–177, September 2008.

[44] T. Suyama, K. Wada, and H. Tanaka. Numerical Simulation of Density Evolution of Dust Aggregates in Protoplanetary Disks. I. Head-on Collisions. *Astrophysical Journal*, 684:1310–1322, September 2008.

[45] M. Takahashi. Zeta potential of microbubbles in aqueous solutions: Electrical properties of the gaswater interface. *The journal of physical chemistry, B*, 109:21858–, 2005.

[46] T. Takahashi. Riming electrification as a charge generation mechanism in thunderstorms. *Journal of the Atmospheric Sciences*, 35:1536–, 1978.

[47] C. Thalmann, J. Carson, M. Janson, M. Goto, S. Egner, M. Feldt, J. Hashimoto, Y. Hayano, T. Henning, K. W. Hodapp, R. Kandori, H. Klahr, T. Kudo, N. Kusakabe, C. Mordasini, J.-I. Morino, H. Suto, R. Suzuki, and M. Tamura. Discovery of the Coldest Imaged Companion of a Sun-like Star. *Astrophysical Journal Letters*, 707:L123–L127, December 2009.

[48] H. Tsuchiya, T. Enoto, S. Yamada, T. Yuasa, M. Kawaharada, T. Kitaguchi, M. Kokubun, H. Kato, M. Okano, S. Nakamura, and K. Makishima. Detection of High-Energy Gamma Rays from Winter Thunderclouds. *Physical Review Letters*, 99(16):165002–+, October 2007.

[49] T. Umebayashi and T. Nakano. Effects of Radionuclides on the Ionization State of Protoplanetary Disks and Dense Cloud Cores. *Astrophysical Journal*, 690:69–81, January 2009.

[50] K. Wada, H. Tanaka, T. Suyama, H. Kimura, and T. Yamamoto. Numerical Simulation of Dust Aggregate Collisions: Compression and Disruption of BCCA Clusters. In *Lunar and Planetary Institute Science Conference Abstracts*, volume 39 of *Lunar and Planetary Institute Science Conference Abstracts*, pages 1545–+, March 2008.

[51] K. Wada, H. Tanaka, T. Suyama, H. Kimura, and T. Yamamoto. Numerical Simulation of Dust Aggregate Collisions. II. Compression and Disruption of Three-Dimensional Aggregates in Head-on Collisions. *Astrophysical Journal*, 677:1296–1308, April 2008.

[52] S. J. Weidnschilling. Production of Chondrules by Lightning in the Solar Nebula? Not So Easy! In *Lunar and Planetary Institute Science Conference Abstracts*, volume 28 of *Lunar and Planetary Inst. Technical Report*, pages 1515–+, March 1997.

[53] E.R. Williams. Sprites, elves, and glow discharge tubes. *Physics Today*, 54:41, November 2001.

In: Lightning: Properties, Formation and Types
Editor: Matthew D. Wood

ISBN: 978-1-61122-198-5

Chapter 6

BALL LIGHTNING AS AN OPTICAL INCOHERENT SPACE SPHERICAL SOLITON*

V. P. Torchigin and A.V. Torchigin
Institute of Informatics Problems of the Russian Academy of Sciences, Moscow, Russia

INTRODUCTION

Ball Lightning: An Unsolved Problem in Atmospheric Physics is the title of the last monograph published in 1999 [1]. We would like to refine this assertion and declare that Ball Lightning (BL) is a shame of physics today. At present more than 2000 papers and reports have been published, more than 200 various BL theories linking BL to a variety of physical, chemical and nuclear processes have been proposed but none of these theories seems to have gained a general recognition because they fail to explain all observed characteristics of the phenomenon. A regular appearance of new theories confirms this conclusion, and new theories are no exception. Systematization and classification of theories have been carried out. However neither of them even approximately can explain enigmatic and intriguing BL behaviour, which to a certain degree reminds us of the behaviour of some highly organized matter. Physicists cannot imagine the object, governed by conventional laws of physics, properties of which coincide with BL properties. Explanation of Ball Lightning physics is deadlocked.

In 1838, French scientist Arago published the first survey of twenty reports of ball lightning. After sixteen years of investigation he wrote in 1854 "The Ball Lightnings of which I have cited so many examples, and which are so remarkable, first for slowness and uncertainty of their movements, and next for the extent of the damage which they cause in exploding, appear to me to be one of the most inexplicable problems of physics today. These balls or globes of fire seem to be agglomerations of ponderable substances, strongly

* A version of this chapter was also published in *Handbook of Solitons,* edited by S. P.Lang and Salim H. Bedore, published by Nova Science Publishers, Inc. It was submitted for appropriate modifications in an effort to encourage wider dissemination of research.

impregnated with the matter of lightning. How are such agglomerations formed? In what region do they originate? From whence are delivered the substances of which they are composed? What is their nature? Why do they sometimes pause for some time in their course, and afterwards they rush off with great rapidity? In the face of these questions, science remains silent". At present, after the passing of 154 years, scientists can answer none of the questions posed at that time. Moreover, prospects of a solution of BL nature will remain foggy if new ideas are not offered. It is extremely difficult to propose them because all conceivable hypotheses already proposed have been considered and recognized as inconsistent.

As noted by Robert Matthews, a reporter for the widely-known journal New Scientist, a Ball Lightning is a flying bag of paradoxes. "It glows like a 100-watt light bulb, yet has no obvious power supply. People say it is linked to thunderstorms but it has rarely been linked to a lightning strike. It emits little heat, yet can melt a glass window. It floats like a ball of gas, yet hangs together like a blob of liquid. To have any hope of explaining the mystery of ball lightning, you must first square all these circles" [2]. Really, it is impossible to imagine an independent isolated object which could feel and bypass obstacles, penetrate into a room through window panes and narrow cracks, changing thus its form, could move against a wind, accompany flying airplanes and not be blown off by a stream of air whose speed is greater than that found in the strongest hurricane. This object radiates cold white light that does not burn down, unlike the white light radiated by any body heated up to several thousand degrees.

A criterion which enables us to evaluate various hypotheses is very simple. Really, as BL can penetrate glass, BL should consist of substances which can penetrate glass. Such substance cannot consist of particles, as any particles (electrons, ions, atoms, molecules, clusters, motes, etc.) cannot penetrate glass. We can conclude that BL has no weight because only particles are ponderable. The same conclusion can be derived from the fact that BL can move against a wind. One does not have to be a prominent physicist to understand that any particles should be carried away by the wind in spite of any inner processes within BL. According to this criterion almost all existing hypotheses should be rejected. The exception is not provided by the most popular present-day BL theory of Abrahamson and Dinnis [3]. They proposed that BL is due to oxidation of silicon nano-particles in the atmosphere. The silicon nano-particles are formed as a result of a reaction of silicon oxides and carbon in the soil during a lightning strike. Exceptions are the hypotheses in which BL occurrence is connected with a presence of external fields and BL represents some discharge in the place where a concentration of these fields takes place [4]. However these hypotheses failed to explain BL behaviour which is determined by the objects surrounding it rather than hypothetical external fields.

Electromagnetic radiation satisfies, first of all, this criterion. Such radiation in a form of light waves can penetrate windowpanes. Windows are fabricated specially from glass which provides a penetration of light into a room. Radiation is not blown off by a wind. The light beam propagates rectilinearly in an atmosphere. It can be bent in the direction where the refractive index of an optical medium increases rather than in the direction of wind. This explains occurrence of mirages in deserts. Speed of light is extremely great. Because of this, light is indifferent to the speed at which an object moves. For example, the speed of an airplane is slower than that of light by a factor of one million.

It is impossible to imagine that light radiation can be concentrated in some limited volumes whose sizes are equal to BL ones without the help of any particles. It is true, but ball lightning exists in air in the atmosphere where molecules of air are in excess. It turns out that it is possible in this case. To avoid immediate rejection of this assertion it is worthwhile to tell a story about how this conclusion was derived. Having prepared an application for a grant, we considered ways to protect optical resonators of the whispering gallery type from the harmful influence that moisture in the air has on their surface that leads to an appreciable decrease in their quality. Such a resonator represents a glass ball of a fraction millimeter in diameter in which the resonance for a traveling wave of whispering gallery type takes place. Such a wave can be imagined as a light wave which circulates inside the ball along its equator. The wave cannot leave the ball because of the phenomenon of total internal reflection on the border between glass and air. Such resonators are characterized by extremely high quality factor Q [5].

The quality factor of such a resonator in the first minutes after manufacturing is equal to about $Q=10^{10}$, and decreases by a factor of 1–2 orders of magnitude in several hours because of the degradation of the ball surface caused by the harmful influence of moisture in the air surrounding the ball. It is known how this phenomenon can be overcome. A similar problem existed in fibers widely used now in telecommunication for high-speed transmission of optical signals at great distances. A fiber was initially a glass thread along which a light, entering into its end face, could be transmitted. The light does not leave the fiber because of the phenomenon of total internal reflection on the border between glass and air. However there was a degradation of the fiber surface in time. Now fibers consist of a glass core where a light propagates. The core is covered by a glass coating which protects the core from degradation. The same phenomenon of the total internal reflection on the border between the coating and core takes place if the refractive index of the coating is a little bit less than that of the core. For fibers, the difference between these parameters can be less than 1 %. Using this recipe we considered a glass ball as a core. We surrounded it in our imagination with a spherical glass coating whose refractive index is smaller than that of the ball. Having calculated a necessary minimal difference in refractive indexes of the ball and the coating surrounding it, we had found out that this difference can be less than 1% for a ball of several centimeters in diameter.

As the field of a light wave is concentrated near a ball's surface, it is sufficient to fabricate in a homogeneous glass a spherical layer of several centimeters in diameter and several microns in thickness where the refractive index is greater by 1 % than that of the homogeneous glass. The layer will confine the light entering into it in the same way as a light is confined by a fiber core. There is an unresolved problem how the light can be introduced into the layer. Unlike fiber, where an end face can be used for this purpose, such end face is absent in the spherical layer.

Then we had remembered that there are optical spatial solitons in nature. Such a soliton can be imagined as a rectangular beam of a monochrome light propagating in a homogeneous nonlinear optical medium whose refractive index increases with an increase in the light intensity. The thickness w of the beam is comparable with the light wavelength, and the width of the beam is close to infinity. In the process of propagation of the beam, its thickness w tends to a steady state and a distribution of the light intensity along the thickness is proportional $\mathrm{Ch}^{-2}(x/w)$ where the x-axis is parallel to the direction along which the thickness is measured. It has been shown that the steady state is stable. At a deviation of w from its steady-state w comes back to a steady state in the process of propagation of the beam. Unlike a usual light beam, diffractive divergence is completely absent in the soliton. It is possible to tell that an intensive plane light beam, propagating in a homogeneous nonlinear optical medium, forms a planar optical waveguide which confines the beam divergence. Thus an optical spatial soliton can be considered as a planar optical lightguide formed by a plane light beam in a homogeneous nonlinear optical medium [6].

As for incoherent spatial optical solitons, they are a generalization of coherent ones. Unlike conventional coherent solitons, incoherent spatial optical solitons are formed by a plane beam of incoherent light. However an incoherent soliton also forms a planar optical lightguide which provides steady-state width of the incoherent light beam. An interest in incoherent optical spatial solitons has arisen rather recently at the end of the 20 century, especially after experimental acknowledgements of their existence [7, 8, 9]. Like conventional optical space solitons, an incoherent optical space soliton can be considered as a planar waveguide whose refractive index is greater than that of the surrounding medium. In this case, an intense light forms a planar waveguide which confines the same light.

Any "lightguide" can guide light as it follows from the literal understanding of this word. For example, a fiber is an optical lightguide which guides a light along its axis. If a fiber is rolled on a finger, a light will propagate along the fiber, i.e. along a spiral trajectory. Similarly, if a planar optical lightguide fabricated in a homogeneous optical linear medium by means of an increase in the refractive index in a planar layer of w thickness is bent, the light will continue to propagate in the bent lightguide along a curved trajectory. Similarly, if a planar optical lightguide fabricated in a homogeneous optical nonlinear medium by means of an intense planar light beam of w thickness (space soliton) is bent, the light will continue to propagate in the bent lightguide along a curved trajectory.

It is clear that the bent lightguide formed by the intense bent light beam is instable because the beam tends to propagate rectilinearly rather than along the bent waveguide and the beam will change the lightguide curvature. But in this case the following interesting theoretical problem arises. What would happen if the curvature of a plane spatial soliton becomes distinct from zero in the whole layer? As far as the soliton form is concerned, an answer is simple enough. If the curvature has been changed along one direction only in parallel with the direction of light propagation, the plane soliton is transformed into a cylindrical one. If the curvature is changed in two mutually perpendicular directions, the plane soliton is transformed into a spherical one. There were strong doubts about stability of such incoherent optical space spherical soliton (IOSSS). Besides, it is not clearly absolutely how initial conditions in a homogeneous nonlinear optical medium can be obtained. Because of this, for some time we believed that results of similar researches could not be checked experimentally and, therefore, it is not meaningful to be engaged in such research.

Later an idea emerged that if IOSSSs can exist, probably they could be observed in nature. Thus attention was paid to Ball Lightnings (BL). As follows from their name, their form is spherical and they are connected with white light as they are visible owing to the light radiated by them. Conventional atmosphere air can be considered as a nonlinear optical medium. Indeed, the air refractive index increases with increasing the air density which in turn is proportional to the air pressure. Due to the optical phenomenon of the electrostriction pressure, an intense light propagating in air exerts additional air pressure which is proportional to the light intensity. As a result, the air refractive index increases with an increase in the light intensity and, therefore, the air is a nonlinear optical medium. In this case an IOSSS can be imagined as a light bubble resembling a soap bubble. However instead of a soap film, there is a thin film of strongly compressed air. The intensive white light circulates in the film in all possible directions and produces the excess air pressure in the film due to the electrostrictive effect. The refractive index of the film is greater than that of the surrounding space. The film manifests itself as a planar lightguide whose curvature is different from zero. The light, introduced into such a film, can propagate within it in all possible directions but it remains in the film and is not emitted into surrounding space. Thus, IOSSS is a combination of an intense light and a film of air compressed by the light. The compressed air provides a circulation of the light in the film. The circulating light, in its turn, provides the air compression. It can be said that IOSSS is a self-confined light circulating in a thin spherical layer of compressed air which is produced by the same light. In other words, the light circulates in a spherical trap, which was set up by the same light. In this paragraph the physical nature of IOSSS in conventional air atmosphere is described completely.

We in the first place have tried to investigate theoretically IOSSS behaviour in the usual terrestrial atmosphere. To our great surprise IOSSS behaviour has coincided with intriguing and mysterious BL behaviour. From here on, the abbreviation BL will be used for a designation of real Ball Lightnings whose properties are known from numerous evidence provided by eyewitnesses, and IOSSS for a designation of abstract light bubbles which exist in our imagination only and whose properties can be investigated theoretically. Since in accordance with our approach, Ball Lightning is a self-confined light of spherical form, we can cut a tail in the word Lightning and name Ball Lightning by the term Ball Light. Thus, IOSSS and light bubble are synonyms as well as BL, Ball Lightning and Ball Light.

In the first chapter IOSSS behaviour in real conditions of earth atmosphere is considered. It is shown that in accordance with well-known physical and optical laws IOSSS behavior exactly corresponds to all features of BL. The authors hope that after reading this chapter a reader will have no doubts about the validity of the hypothesis because too many BL features can be explained in a perfectly natural way. All of these explanations are collected in one place in order to convince a reader of the validity of the proposed approach.

It would be possible to finish studying an interrelation between BL and IOSSS by the first chapter and to wait until the presented hypothesis wins a general recognition. But there was a unique situation for the authors. Unlike others, they knew that IOSSSs do exist in nature. It would be silly to ignore this advantage. First, results of two-centuries of investigations involving attempts to produce an artificial BL in a laboratory have been analyzed. It turned out that various luminous objects have been produced by many experimenters. In the 19th century, similar objects did not exist autonomously. They disappeared at once after ceasing a gas discharge. In the 20th century, experimenters succeeded in producing objects which can exist autonomously for several seconds. At the present time, similar objects can exist about 10 seconds after ceasing a gas discharge. This time is comparable with the lifetime of many BLs. Investigators differently name the objects observed by them. Among these names were terms such as long-lived luminescent objects, ball-shaped plasma formations, exotic vacuum objects, spherical anomalous objects, luminescent ball-shaped objects, abnormal anomalous objects, abnormal plasma formation etc. We will name shortly these objects as anomalous objects (AO). On the assumption that AOs are miniature IOSSSs with a relatively small lifetime and little stored energy, we succeeded in explaining AO behavior in many experiments. Results of these studies are presented in [10]. Unlike BLs, whose properties are accessible from evidence of eyewitnesses only because it is a seldom enough natural phenomenon, AO properties can be studied carefully in a laboratory. An analysis of these properties on the assumption AO is OISSS enabled us to discover and justify several new optical effects. Firstly, an analysis of AO spectrums prompted that along with Kerr-like and electrostriction nonlinearities, there is a new type of nonlinearity in nature. The nonlinearity takes place in gas mixtures and manifests itself by adsorption of molecules of the mixture component with maximal refractive index in the field where an intense light beam is propagating [11, 12]. This type of nonlinearity is stronger essentially than that of nonlinearities mentioned above.

An analysis of possible mechanisms responsible for AO appearance enabled us to conclude that an intense light in a gas is subjected to self-organization [13]. As a saturated vapor with a decrease in its temperature is converted into a mist consisting of drops of liquid, an intense light and gas are converted into a majority of AOs with an increase in the light intensity. This effect changes radically Kirchhoff's notions about properties of equilibrium light radiation. At great enough light intensity in a gas, a state of equilibrium is impossible because of a self-organization of the intense light. Since the BL nature became known, attempts had been undertaken to consider mechanisms of its appearance. Results of these studies are presented in [14, 15].

Thus, having accepted a priori, that IOSSS exist in nature, and having learned in theory their behaviour in accordance with conventional physics laws, we were able to conclude that IOSSS properties and behaviour peculiarities are similar to BL and AO ones. This gave solid grounds to argue that IOSSSs exist in nature as a matter of fact.

It was a much more difficult task to justify IOSSS existence theoretically. Apparently, such a conclusion appears extravagant. Indeed, it is very hard to believe that a conventional white light can be self-confined by usual air in such a way that it is circulating in a shell of several centimeters in diameter. Common sense protests against it. The authors were forced to overcome a series of irrefutable and indisputable, at first sight, arguments that the IOSSS cannot exist in principle. They were forced to justify the possibility of light circulating into the air along trajectories with a small curvature radius [16], stability of IOSSS [17], and their extraordinarily large lifetime [18]. The nature of electrostriction pressure exerted by intense light in gases has been investigated anew in detail.

At present many physicists are very interested in this approach. More than a dozen scientific peer review papers have been published in international physics journals such as Doclady Physics, Physics Letters A, Optics Communication, European Physical. This activity has been carried out in a frame of conventional physical notions and laws in conditions when a positive answer is known in advance and it is only necessary to find a solution. Certainly, the confidence in a positive final outcome of researches has promoted their success. The authors had no purpose in elaborating a model of ball lightning that could be used as a basic argument confirming the validities of the chosen approach because any model is a fine target for critique. Indeed, in any model it is impossible to take into consideration all varieties of phenomena in the real world. The task was more modest. It was necessary to disprove major objections which were expressed by physicists most often. The basic argument confirmed a validity of the approach is not theoretical conclusions, but the dramatic coincidence of BL and IOSSS properties.

After there was confidence that IOSSS exist in nature, there was a lawful question—whether there can be similar IOSSSs in other phenomena accompanied by wave processes. Since the energy stored in IOSSS is gradually decreasing due to inevitable radiating losses inherent in waves of the whispering gallery type, the IOSSS lifetime is definite. IOSSS ceases its existence abruptly. A similar situation takes place for nuclei of chemical elements which, as is known, are subjected to a radioactive decay. In this connection, it is interesting to consider a hypothesis where a nucleus is presented by a certain space spherical soliton (SSS). As is known, a nucleus is described by a wave function which satisfies a wave equation. Like incoherent optical SSS where light waves of whispering gallery type are circulating, waves of the same type but presented by a wave function can circulate in a nucleus. In this case a nucleus can be imagined as an oscillator with an extremely high quality factor. A very valuable property of such an oscillator is its ability to store and accumulate an action produced by other similar oscillators. In principle, such ability permits one to obtain a noticeable result in the process of a week but prolonged interaction between nuclei. In the next section, properties of a system consisting of a majority of identical oscillators interacting one with other are considered. On the basis of rather general assumptions conditions are found, at which a weak but prolonged interaction can result in a noticeable change in parameters of some oscillators. These results enabled the formulation of conditions favorable for an increase in the speed of radioactive decay. Numerous experimental data confirming these results are presented.

Five years ago the authors knew nothing either about Ball Lightnings, or about radioactive decay. Having shown curiosity and asking a simple question—whether there can be self-confined light in the form of a light bubble in a homogeneous nonlinear optical medium—the authors have plunged into the world of the abnormal phenomena from information collected over centuries. It has appeared that many of these phenomena can be explained with attraction of the concept of light bubbles. Researches have been led by the personal initiative of the authors. All ideas and conclusions reflect their personal opinion.

BEHAVIOR OF OPTICAL INCOHERENT SPACE SPHERICAL SOLITONS IN THE ATMOSPHERE

Abstract

On the assumption that optical incoherent space spherical solitons in a form of light bubbles can exist and their shell, unlike soap bubbles, consists of strongly compressed air where an intense white light is circulating in all possible directions, the behavior of such bubbles in the conventional air atmosphere is considered. It is shown that a bubble, like a usual light beam, should be deflected in the direction of the air refractive index gradient where the bubble is located. Distributions of the gradient in various situations which took place during observations of Ball Lightnings are considered. It is shown that the distribution of the gradient in all these cases is so that a simple and natural explanation of Ball Lightning behavior near the earth's surface, moving in a horizontal direction, moving against a wind, penetration in rooms and so on is possible. Thus, the physical parameter which controls a Ball Lightning motion is discovered. The authors hope that a reader loses all doubts after reading this chapter that there are light bubbles in nature and Ball Lightnings are their experimental confirmation.

Taking into account that BL represents a light bubble (LB), consisting of a thin layer of strongly compressed air (shell) where an intensive light circulates in all possible directions (figure 1), we shall consider behaviour of such bubble in the conventional air atmosphere.

The motionless LB in a homogeneous atmosphere remains motionless. Really, the central symmetry of the system in this case takes place, and there is no selected direction. If LB is located in an inhomogeneous atmosphere where the refractive index *n* depends on spatial coordinates, the situation varies. As the light beam, propagating in an inhomogeneous medium bends in the direction of the gradient of the refractive index, any closed light beam circulating within the LB shell tends to be shifted in the same direction. In other words, LB tends to move where the density of air is greater. LB is the most sensitive device of inhomogeneity of the air density (provided that other parameters of air in all points of considered space are homogeneous). Really, under the assumption that the light circulating in LB shifts per one revolution only at 1 micrometer, its shift per one second is about one kilometer. It is explained by the extremely great light speed of 3 10^8 km/s. For example, if the length of LB circumference is equal to 30 cm, light will make one billion revolutions per one second. A billion shifts by one micrometer makes a total shift in 1 km.

There are no BL theories which can explain an extremely simple thing. What are the reasons that force BL to move in such a way that BL moves? For example, we can read in [1]. “In the author’s model electrostatic repulsion is invoked partly to explain why balls falling from the cloud base and therefore obviously heavier than air, so frequently stop before they hit the ground”. Our explanation is very simple and natural. There is no gravity attraction to the ground, or to be more precise, forces of gravitation are inessential. As is known, the air pressure decreases with an increase in the height. Therefore, the air density and air refractive index decrease with an increase in the height. Since BL moves along the gradient of the refractive index, BL moves downwards to the ground surface until the above-mentioned dependence of the refractive index on the height is held. But BL does not achieve the ground surface. The reason is in the temperature of the ground surface which absorbs the sun’s radiation. Because of heating, the temperature of the ground surface is greater than that of the air near it. Because of the air heat conduction, the air temperature of a layer near the ground surface is greater than that of layers residing at greater distance from the ground surface. Since the air density is in inverse proportion with the air temperature, the air density near the ground surface is lower than that of the layer located at greater distance from the ground surface. As a result, there is the maximum of the air density at some height h. This height is about 1 m. Achieving this maximum, BL stops motion along the height and moves along the horizontal component of the air density gradient or the air refractive index gradient (directions of these gradients coincide). That is why observable BLs move mainly at a low height in a horizontal direction. Readers should remember that antifog headlights on cars try to be as close to the earth’s surface as possible. The air temperature near the earth’s surface is maximal and fog is not formed at such temperatures. Therefore low located antifog headlights provide light beams penetrating at a greater distance.

Moving in a horizontal direction, BL itself can change distribution of the air density around itself, especially at approach to obstacles. If air pressure is homogeneous, the air density is in inverse proportion to the air temperature, therefore BL moves in the direction opposite to the temperature gradient. Coming nearer to some obstacle, BL heats up due to radiation. In turn, the obstacle heats up the air surrounding it and thus decreases the air refractive index. Gradient of the refractive index directed from the obstacle appears. Moving along this gradient, BL jumps aside from the obstacle or bypasses it. Externally it looks like as if BL would "feel" the obstacle and bypass it.

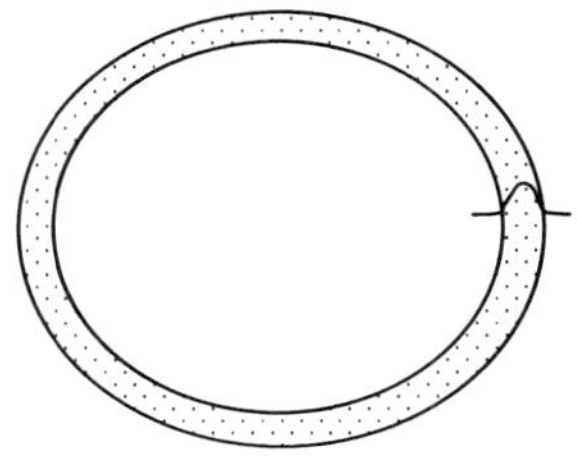

Figure 1. Light bubble structure

Figure 1. Light bubble structure

As is known, falling BLs are observed approximately 10 times more often than rising ones [2]. It speaks to the fact that the gradient of the refractive is directed downward in a typical atmosphere. However air temperature can be different in different places of space and, hence, different air density at the same pressure is possible. In particular, an inverse distribution of the air density along the height is possible where BL will move upwards.

How a Soliton Finds Holes, Splits, Chimneys to Penetrate

There is a lot of evidence that BLs can penetrate into rooms through small cracks and splits. We present some of them taken from the Stakhanov book [3].

During a strong thunderstorm, BL of 20–30 cm in a diameter entered a room through a hole for grounding in a wall.

During a thunderstorm, BL of 10 cm diameter passed in a hole of 2 cm width, "having taken form of a sausage".

BL penetrated through a crack in a windowpane. BL diameter was about 10–20 cm.

BL of 30–50cm diameter entered through a small hole in a window (the broken off corner of glass) of 1–1.5 cm width in a form of "a yellow thread". Having done some circles around a room, it is blown up in 20–30 s.

BL of 5–10 cm diameter has entered "by snake" during a thunderstorm through the slightly opened window leaf, having formed then a ball. Having passed in a room a distance of 5–10 m, it disappeared without explosion near a switch.

During a strong thunderstorm, BL has entered into a house in a split between boards near a chimney tube, thus the board started smoking, and there was a fire.

BL of 10–20 cm diameter has passed through a hole of 8 cm diameter.

BL of "tennis-ball size" has passed through the closed window with a crack in the glass.

BL "flows" in a hole between logs in a smithy. The width of the hole was much less than BL size. In the smithy it looked like a sphere of 12–13 cm diameter of orange color and shone as a 50–100 W lamp.

In a crack in a wall a yellow ball the size of a large orange squeezed through. More truly, it did not squeeze, but it was poured from one half to another.

BL has passed in a room in a hole in glass, having flattened as its size was greater than the hole. Eyewitnesses specify: "The sphere was 10–15 cm from our faces, and we well saw, how it began to pass through a hole, taking the form of a melon. It was extended, became smaller in diameter and has passed through the hole. When the sphere was passing through the hole and was decreasing in its size, it was as though it was shivering all the time, and it seemed that it consists of galley."

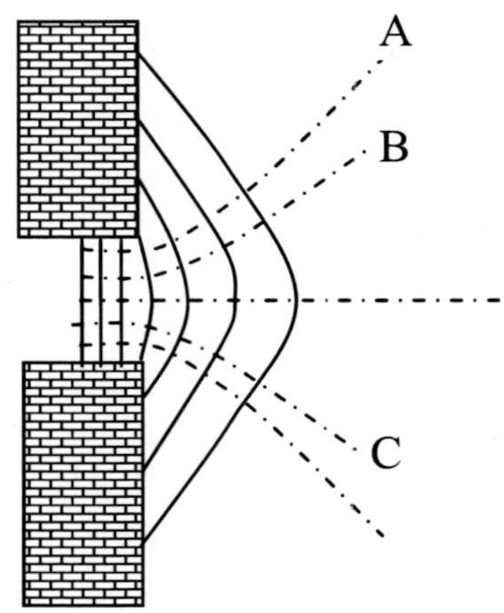

Figure 2. Distribution of the gradient of the refractive index near a tunnel

Before analysis of the mechanisms providing penetration for BL through small cracks and holes, we should answer a simpler question—how does a soliton find similar objects? In general, why isn't a soliton indifferent to cracks and holes? Clearly the soliton has no organs of sense, vision, sense of touch that are found in a living being. However the soliton can "feel" the slightest changes in the air refractive index and moves along the refractive index gradient. We shall show that cracks are sources of such changes.

Consider a soliton movement in a tunnel shown in figure 2 between a room (left side in figure) and external space (right side). Let the temperature indoors be smaller than that outdoors, and the air pressure indoors and outdoors are identical. In this case the indoor air density is greater than that outdoors, and the gradient of the refractive index at the tunnel axis is directed to the room. As the air density along the tunnel changes gradually and continuously, surfaces of identical density look like solid lines shown in figure 2. Dashed lines in the figure are perpendicular to these solid lines. The direction of a tangent to any of these dashed lines in any point coincides with a direction of the refractive index gradient in this point. It is easy to be convinced, that a soliton, moving along any dashed line from various points A, B, C, moves in a direction to the tunnel. A peculiar funnel appears near the tunnel. Having got into the funnel, a soliton moves in a direction to the tunnel.

In the same manner solitons find chimneys on roofs of buildings to penetrate through the chimneys in rooms. There is a funnel near each chimney. Dropping downwards, BL meets one such funnel along its way and begins its motion as is described in the next section. We ought to note the summary area of cross sections of all funnels on any roof is significantly greater than the summary area of cross sections of all chimneys. This is why a probability to penetrate into a chimney is great enough. At least, the probability is greater than that for rain drops because there are no funnels for them.

How Soliton Penetrates a Room through a Small Split

It is interesting also how a soliton determines the size of the tunnel cross-section (hole) and deforms itself in such a way that it can penetrate the hole. To answer the question, one ought to consider the reasons that provide a spherical form for the soliton in a homogeneous air and deform the soliton in an inhomogeneous air. The air pressure within the soliton is great enough. But this does not mean that the soliton is similar to a child's air balloon which can not be deformed essentially. The air pressure in the soliton exists in a soliton shell only which is a thin layer of several micrometers width. The pressure provides a spherical form for the soliton, but in much smaller degree than the pressure within the whole ball volume as in the child's air balloon. This entails that external small forces can deform the soliton significantly. The soliton can be compared with the formless shell of a child's air balloon located in weightlessness and a vacuum. The shell takes the form of a ball because its own weak elastic re-established forces of the shell can show themselves in the absence of other strong forces of gravity and the air pressure. These re-established forces tend to straighten the shell. But the shell can not be a plane because of its closed form. As a result, the shell ends up as a ball. But this ball is different from a child's air pumped balloon. For example, such a balloon can not be drawn through a small ring. The airless shell in the weightlessness with the same spherical form can be equally well drawn through the small ring as the airless shell can be drawn through the small ring in the presence of gravity.

A similar situation takes place for the soliton shell which takes a spherical form in a homogeneous optical medium owing to elastic properties of a layer of air compressed in the soliton shell. A soliton form can be easily changed in precisely the same way as the spherical form of an elastic shell of a balloon is changed in weightlessness and a vacuum.

As the soliton approaches the hole, inhomogeneity of the air density arises because of heating of the air by the soliton. But the soliton heats the air indirectly. Non-transparent objects should be heated before. Such an object is a region around the hole shown in Figure 3a. The soliton heats the region near the hole denoted by the letter *H* in Figure 3a. The region heats the adjacent air on account of the heat conduction phenomena. The nearer the air region the non-transparent surface is, the greater the air temperature. As a result of the heating the air density decreases and forces come into existence that push the soliton surface from the hot region. On the other hand, there is the air refractive index gradient near the hole axis directed inside the hole. As a result, there are forces near the hole axis that pull the nearest regions of the soliton surface into the hole. These opposite forces entail soliton deformation because the re-established forces of the soliton are weak. The soliton form is shown in figure 3b. As the soliton penetrates the hole, the following feedback comes into force. The smaller the distance between the soliton and the side surface of the tunnel the greater the heating and the greater repulsion of LB from the side surface. Thus, the deformed soliton is located in the middle of the tunnel cross-section as is shown in figure 3c. The next steps of the soliton penetration are shown in figures 3d, 3e, and 3f.

Thus, the forces connected with the initial gradient of the air density draw the soliton gradually through the tunnel. The soliton resumes its initial form after the soliton leaves the tunnel at the opposite side of the wall as is shown in figure 3g. One can say that the forces that are responsible for passing the soliton through the tunnel and forces that are responsible for movement of the soliton along the gradient have the same origin. But because of heating, the new effect of self-action takes place. The soliton heats the air in such a way that the soliton decreases its cross-section and floats through the tunnel in the middle of the hole exactly without touching its side surface. Since the soliton shell thickness is about several micrometers, a soliton can penetrate through very thin slots of about ten micrometers width.

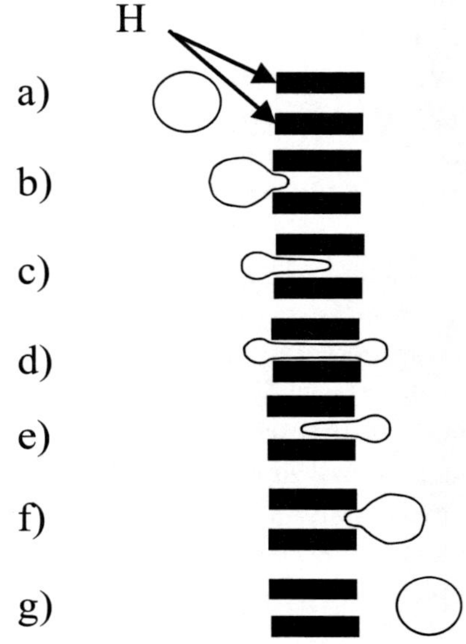

Figure 3. Steps of penetrating BL through a hole in a wall

How Soliton Penetrates a Room through Glass Window Panes

Now it is established reliably, that BL can penetrate windowpanes, sometimes destroying them, but in most cases leaving them untouched. Damages, if they are present, have local character and they are in those places through which BL penetrated. In work [4] 42 cases of BL penetration in closed rooms are described. It is necessary to note, that the overwhelming majority of BL theories cannot explain this BL property and for this reason should be rejected. It concerns all theories in which any particles such as electrons, ions, clusters, as well as plasma are taken into account for explanation of BL nature. As is known, any particles cannot penetrate through glass. Glass tubes and flasks used in chemistry are good confirmation of this. But the light penetrates through the glass. Because of this the glass is used in windows. But a part of the light is absorbed by the glass. The glass can be heated if a very large intense light wave is passing through it. This can cause its damage. There are photos of a glass disk of about 8 cm diameter that was displaced from a window glass after BL penetrated it [5].

A soliton finds windows in the same way that it finds cracks in walls because the characters of distribution of the air density near windows and cracks are the same. The gradient of the air refractive index is directed to the room in both cases. If the temperature in a room is smaller than outside, the heat penetrates the room through a windowpane. It is known, that the heat moves in the direction opposite to the temperature gradient. It means that the temperature gradient is directed outdoors from the room. The gradient of the air density is opposite to the temperature gradient and, hence, it is directed to the room. The character of distribution of the air density near a crack shown in figure 2 can be applied also to the window. In this case, the crack can be considered as a window of the same form with the glass thickness equal to zero. Because of this, a soliton finds windows in the same way as it finds holes.

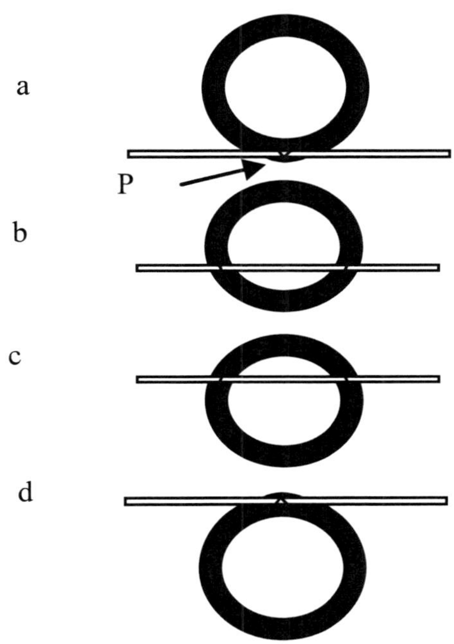

Figure 4. Steps of penetrating BL through a plane glass

A soliton penetrating a window glass appears as follows. As is known, a light beam, incident at a plane glass plate, always passes through it. In this case the incident angle α is equal to the angle β at which the beam leaves the plate at its opposite side. The light beam propagates seemingly reclining and cannot return backwards to the soliton. But this is true only if the refractive indexes are the same at both sides of the glass plate. The refractive index within the soliton is greater than that at the opposite side of the plate where the air pressure is equal to that of the atmosphere. In this case $\alpha<\beta$ and there is the total inner reflection from the outdoor glass-air boundary at $\alpha\cong 90^0$. Thus, two regimes are possible. The light passes through the glass plate in the first regime and totally reflects in the second one.

When the soliton touches the glass at $\alpha\cong 90^0$, the light penetrates in the plate but the light does not leave it at the opposite side of the glass. The light reflects from the boundary: air-opposite side of the glass on account of the phenomenon of total inner reflection, that is where there is a second regime. After the reflection the light enters the soliton shell again. In addition, there is an intense light field in the air near the boundary. The field forms a region near the boundary with the excess air pressure as is shown in figure 4a by letter *P*. As a result, the next portions of light penetrate in the region and it increases in volume as is shown in figure 4b. In this case there is the first regime at the air-glass boundary.

If the densities of atmosphere air are different at the opposite sides of the window glass, then in accordance with the described above effect the soliton moves in that side where the air density is greater. There is an impression that the soliton penetrates unobstructed through the glass. In fact, only light penetrates unobstructed through the glass and forms a layer of the compressed air at the opposite side of the glass as is shown in figures 4c, 4d. The compressed air confined by the initial soliton does not penetrate through the glass.

OTHER PUZZLES OF SOLITON MOTION

Usually the homogeneity of air parameters indoors is smaller than that outdoors because air inside a room is mixed to a smaller degree than outdoors. Since warm air is lighter than cool air, the air near a floor is cooler than that near a ceiling and therefore the air refractive index is maximal near the floor. There is much evidence that BL, having gotten into a room, moves near the floor, but there is no information showing that BL rises to the ceiling.

There is no problem explaining why a soliton bypasses obstacles. Approaching an obstacle, the soliton heats it due to light radiation. In turn, the obstruction heats a layer of air near it owning the heat conduction and convection phenomena. As a result, the air density near the obstacle decreases and the soliton is repelled from it. In this case a BL, appearing in a room, moves towards the floor which presents itself as an obstruction. Besides, the walls of the room are cooler than the air in the middle of the room. That is why there is a stove with a chimney in the room. The stove heats air near it and after that the warm air heats the walls. Approaching a wall, the soliton heats air between the wall and soliton. As a result, its motion towards the wall ceases and the soliton can move along the wall. Since the air before the soliton is cooler than that behind it, the soliton continues to move along the room perimeter at some distance from walls. In some cases the soliton can find a door, penetrate though it and leave the room. There is much evidence supporting this picture.

For example, I. Stakhanov [3] describes in his occasion 29 a closed trajectory BL in a carriage where builders lived. The trajectory is shown in figure 5. Since there is a thermal contact between the walls and the ground, the temperature of the wall is lower than that of the air in the room. In this case the air temperature near the wall is lower that that in the middle of the room and the soliton tends to go toward the nearest wall. The soliton heats the air near walls and resides at some distance from the wall. Since the air temperature is greater behind the soliton than in front of it, the soliton moves along the wall. Meeting a perpendicular wall which prevents its movement in a straight line, the soliton heats the wall. As a result, the soliton turns 90 degrees. Achieving its entry point and heating the air in the room, the soliton “sees” that the outdoor air is cooler than that in the room. In this case the soliton leaves the room. In some cases the soliton can perform several bypasses along the room perimeter (Stakhanov, evidence number 45). By the way, it is marked in some evidence that some signs of burning on walls has been seen.

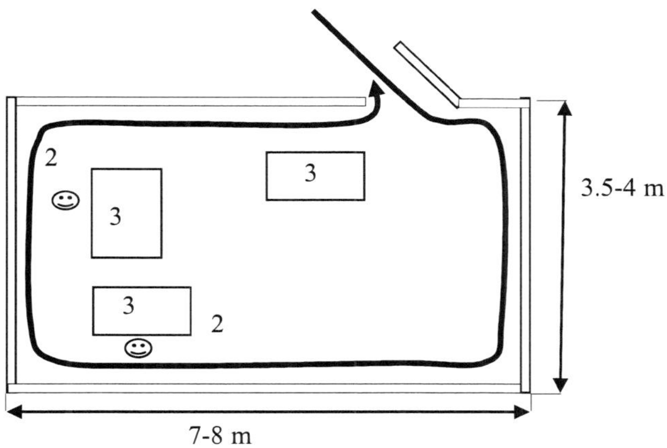

Figure 5. Trajectory of Ball Lightning in a carriage

There is much evidence that BL moves not only at an unchanged distance from a wall or floor but also by bounces. It can be explained as follows. We supposed above that the time of transient processes is negligible, and there is a steady-state constantly. If there is some latency, for example, the walls or floor are wet and their heating requires some time, an oscillating process can take place. The soliton approaches a wet floor at the distance which is smaller than that for a dry floor because the feedback comes into play with some delay. In this case the air temperature between the soliton and the wet floor is greater and the soliton is repelled from the floor. The amplitude of soliton oscillations depends on a degree of floor wetness. By the way, the same effect takes place at a soliton penetration through a hole. As is reported in evidence number 100 (Stakhanov, 1997) a ball of 10–15 cm in diameter, passing through a small split in a pane, changed its shape and trembled like jelly.

A perfectly different situation takes place when the soliton approaches objects with great heat conductance (metals, water). In this case the soliton is not able to heat the surface of the object because the heat propagates into the object. As a result, the soliton energy decreases, the soliton becomes instable and explodes. The energy liberated from the soliton at the explosion can damage the objects. There are numerous evidences that BLs cease their existence at touching metallic objects (Stakhanov, 1997, evidence 23, 31, 33, 34, 35, 37, 40, 48, 49, 51, 78).

It is a very interesting case where BL touches the face of a long, thin metallic wire whose mass is small and the air is heated by the wire to a small degree. In this case the soliton has had time to heat the wire at the temperature of vaporization. Since the refractive index of metallic vapor at the temperature of vaporization is greater than that of the air at normal conditions, the soliton does not repel from the hot wire but is attracted to it. As a result, the soliton can burn a hole. For example, LB burns a hole of 5 mm in diameter and 3 mm depths in a cleaning rod (Stakhanov 1997, evidences no.4, 34, 35).

The case that the authors learned about from K.K.Petrovskiy is rather indicative in this respect. A Ball Lightning burned out in a cable with a polyethylene shell and a copper core at a distance of several meters, and left the shell almost untouched. This fact can be explained as follows. Penetrating in the hole of the coating, the soliton acquires the shape of a sausage shown in figure 6. Dependence of the gas density on the distance along the coating hole is also shown in the figure. As is seen, the gas density decreases with an increase in the distance from the face of the wire because the gas pressure decreases. Indeed, the pressure in part of region 1 is greater than that in part of region 2 located at a greater distance from the face than region 1 because the gas moves from region 1 to region 2. The gas refractive index is proportional to the gas pressure, all other conditions being the same. Metal within the shell evaporated so quickly that the heat had no time to extend into all sides of the shell. The greatest pressure and, hence, the greatest refractive index is at the end face of the evaporated wire. Therefore during the entire time of evaporation of the wire, the soliton in the form of a thin sausage is located within the polyethylene shell.

Another occasion took place with the Soviet propeller airplane Li-2 which was attacked by BL at the height of 3300 m in partial rain clouds. Immediately after passing a cold thunder front, the airplane was covered by ice. The height of clouds was about 5.5–6 Km and the air temperature was about -2 -4^{0} Celsius. An orange-red fire ball of 25–30 cm in diameter approached the airplane in the front. At the distance of 30–40 cm from the airplane's nose the ball turned to the left and flew past the pilot's cabin. At that moment it touched the right propeller and exploded. A single damage was discovered to a propeller blade. Singer presents a drawing of the damage, which is shown in figure 7.

Since a propeller pulls a plane forward, the excess air pressure takes place near the propeller. The excess pressure is maximal near the end of the propeller blade because the linear speed of this area of the blade is close to maximum. Thus, the pressure is maximal in that part of the propeller blade where damages to the blade edge are seen. Residing in the place of maximal pressure, the soliton rotated with the propeller and burned out craters in the same manner as BL burns out craters in motionless metallic objects. It does not matter that the linear speed of the propeller region where the damage took place is extremely great. The light speed is greater by one million times. The soliton resides in the space where the air density is maximal independently of whether this space is motionless or moves at great speed.

If an obstacle is wet, water vapor appears. Since the refractive index of the water vapor at a normal pressure is equal to n_{water}=1.0002354 and the refractive index of the air at a normal pressure is equal to n_{air}=1.0002727, then the refractive index of the appearing gas mixture decreases not only owing to an increase in the temperature but also owing to insertion into the air of a gas component with a smaller refractive index. Thus, a wet obstruction promotes better repulsion from it.

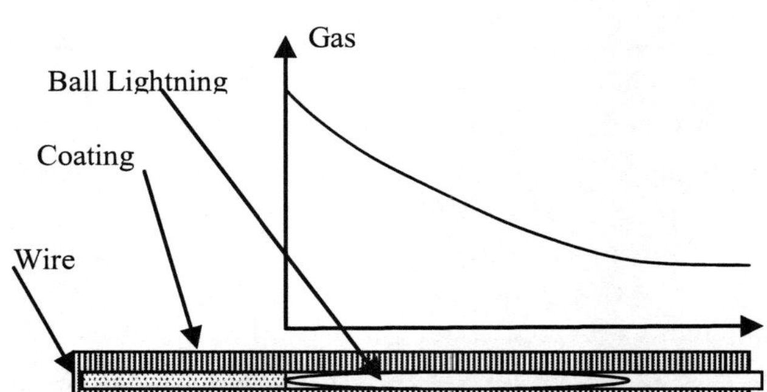

Figure 6. Ball Lightning deformation within the hole where the copper core of TV cable resided and dependence of vapor pressure in the hole on the distance along the cable

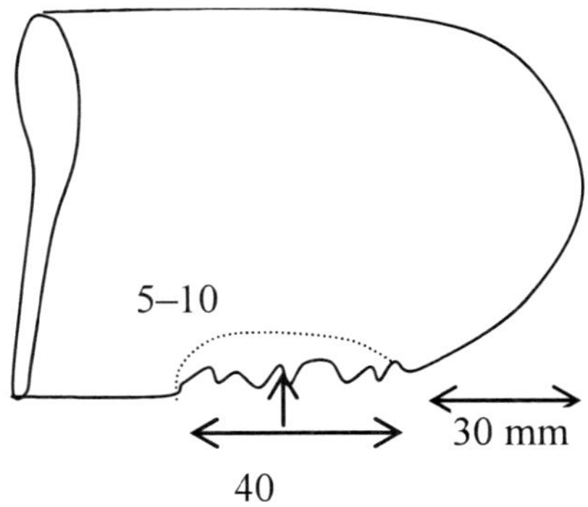

Figure 7. Damage to propeller blade caused by Ball Lightning

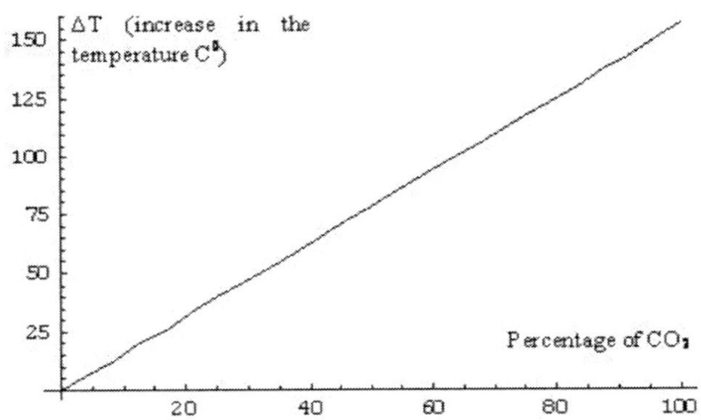

Figure 8. Any point above the incline line shows parameters of CO2 and air mixture whose refractive index is smaller than that of pure air at the temperature 0^0 C

A different situation takes place if CO_2 gas is inserted into the air, as takes place in a burning stove. Since the CO_2 refractive index n_{CO2}=1.0004197 is greater than the air refractive index, the soliton can be drawn into the region where CO_2 percentage is greater. In figure 8, the straight incline line divides a majority of points (ΔT, percentage of CO_2) into two parts. The points above the line correspond to the refractive index of the gas mixture that is smaller than that of the air at normal conditions. The soliton located in the air at normal conditions is repelled from such a gas mixture. On the contrary, the soliton located in the air at normal conditions is attracted to the gas mixtures corresponding to points below the curve. This example shows that BL can be drawn into a region of space where the temperature is greater.

Thus, a soliton is the most sensible device for determination of the inhomogeneity of the air refractive index or air temperature (if air composition is homogeneous and air pressures in various points of considered space are the same). The soliton moves in a homogeneous air in the direction opposite to the gradient of the air temperature. One ought to take into account that, approaching some obstruction, the soliton heats it and the air near it and in such a way changes a distribution of the air temperature in space. As a result the soliton can bypass obstructions.

SPEED OF SOLITON MOTION

Let us estimate preliminarily the velocity at which a light bubble moves in an inhomogeneous atmosphere. For the sake of simplicity, consider an infinite cylindrical layer of compressed air in which an intense light is circulating and supporting the air compression. Let the cylindrical layer of R be located in an inhomogeneous atmosphere whose refractive index increases along the x-axis at the rate $g_n = dn/dx > 0$ and does not depend on y, z coordinates (figure 9). If the cylindrical layer is located in a homogeneous atmosphere where $g_n = 0$, then the light makes one revolution around the cylinder axis in time

$$T = 2\pi Rn/c \tag{1}$$

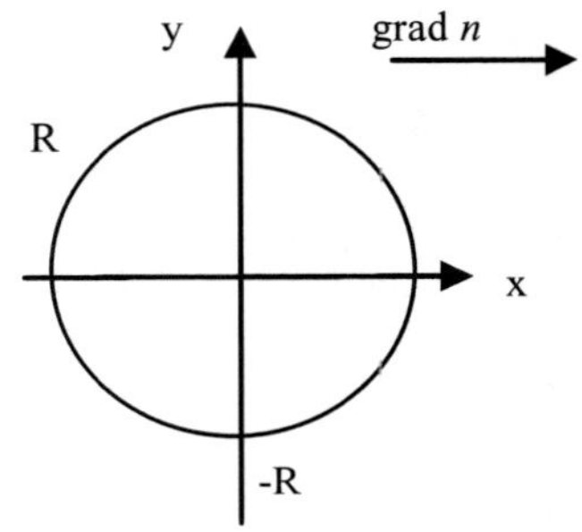

Figure 9. Directions of coordinate axis in infinite cylinder

If the cylindrical layer is located in an inhomogeneous atmosphere where $g_n > 0$, then the light trajectory is not closed. The light shifts additionally at some distance Δx along the x-axis. In accordance with the eikonal equation the additional shift of a light beam in a presence of the additional refractive index gradient g_n is determined by the following expression

$$d^2\Delta x/dt^2 = g_n c^2 \mathrm{Cos}(ct/R) \tag{2}$$

Let the light beam propagate along the arc beginning at point $x=0$, $y=-R$ and ending at point $x=0$, $y=R$. Supposing that at time $t=-(\pi/2)R/c$ the initial speed $d\Delta x/dt$ and initial shift Δx are equal to zero, we obtain from (2) that the total shift of the light beam is

$$\Delta x = (\pi/2) g_n R^2 \tag{3}$$

In this case the speed of shifting is determined by the following expression

$$v = 2\Delta x/T \tag{4}$$

Taking into account (1), we obtain

$$v = g_n Rc/2 = \Delta n_D c/4, \tag{5}$$

where

$$\Delta n_D = g_n D \tag{6}$$

is the change in the atmosphere refractive index at a distance equal to the light bubble diameter. For example, if $\Delta n_D=10^{-5}$, then $v=750$ m/s, that is v surpasses the sound speed in air. We ought to emphasize that a soliton movement is not accompanied by a movement of the air compressed within it. Like light bubble movement through window glass, where only light radiation penetrates the glass and compresses a new air portion located at the opposite side, light bubble movement in the air atmosphere is accompanied by compression of new air portions along the direction of the movement and corresponding release of air portions from the opposite light bubble side.

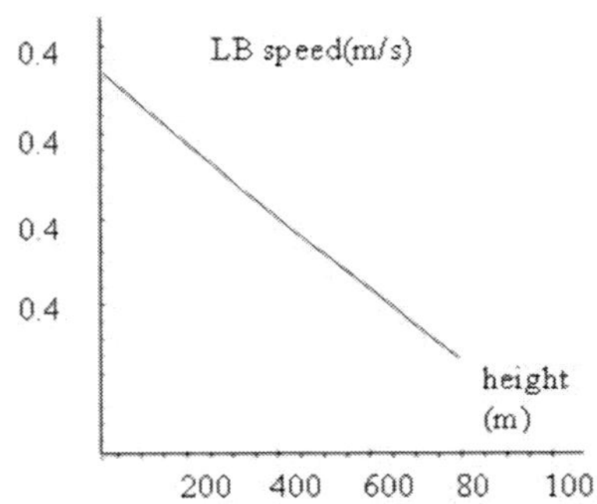

Figure 10. Dependence of light bubble speed on height

As is known, the air pressure and therefore the air refractive index gradient change with the height. Because of this the soliton speed changes with the height, too. Let estimate the soliton speed of falling free. The dependence of the air pressure p on the height h is determined by the following barometric formula

$$p(h) = p_0 \exp(-h/H), \tag{7}$$

where $H=kT/(\mu m_1 g)$, $k=1.380\ 10^{-23}$ J K^{-1}, μ is the molecular mass of the gas (for the air $\mu=29$), m_1 is the mass of the hydrogen atom and $m_1=1.66\ 10^{-27}$ kg, T is the gas temperature (K^0), g=9.8 m/s is the acceleration of falling free; as is seen p_0 is the pressure at $h=0$. At $T=300$ K^0 we have $H=8875$ m. In this case the dependence of the soliton speed on the height is shown in figure 10.

Certainly, there are temporal changes in a local gradient of the refractive index in space and time, which are essentially greater than the constant change connected with constant decrease that shows itself as a lightguide which can guide the light around the earth's surface. In this case the gradient of the air refractive index is greater than the curvature of the earth surface $C=1/R=1.57\ 10^{-7}$ m^{-1} where R is the radius of the earth. Then the soliton speed in the region in such gradient is 2.35 m/s. At the same time there are some regions, for example frontal zones, where the air refractive index gradient is significantly greater and the soliton speed can achieve tens of meters per second. As follows from the statistics, BL falling downwards are observed approximately 10 times more often, than rising upwards [2].

How a Soliton Catches Flying Airplanes

One can conclude that the probability of BL appearance near the earth's surface is greater than that at great heights where airplanes are flying. Why do BLs prefer to visit flying airplanes (there are many reports) and are perfectly indifferent to airplanes on the earth's surface? Why such a yearning for airplanes in flight? Everybody who looked in an illuminator of a flying airplane and saw small clouds and swarms that ran away with high speed may ask why BLs are not blown away from the airplane wing although very strong winds at speeds of about 1000 Km per hour blow on the airplane. If BL were composed of some particles, clusters, ions, electron etc., all of these would be blown away immediately.

To answer these questions let's estimate the degree of atmosphere inhomogeneity produced by a flying airplane. An additional air pressure Δp near the front edge of a wing where the air speed relative to the airplane is maximal and is equal to the airplane speed u relative to the earth is determined by the following expression $\Delta p=\rho u^2/2$, where ρ is the air density and $\rho\cong 1$ kg/m^3. At the airplane speed u=720 Km/h=200m/s we have $\Delta p=2\ 10^4$ Pa≅0.2 atmospheres. The air refractive index at the normal atmosphere pressure is determined by the expression $n=1+\Delta n$, where $\Delta n=2.7\ 10^{-4}$. Then an increase in the air pressure by 0.2 atmosphere entails an increase in the air refractive index by $0.2\Delta n\cong 0.54\ 10^{-4}$. In accordance with (4) to provide a soliton motion at the airplane speed the following air inhomogeneity is necessary: $\Delta n_D=u/(c/4)\cong 2.66\ 10^{-6}$. In this case the soliton may catch up to the airplane located from it at distance $S\cong D(\Delta n/\Delta n_D)\cong 20D$ where D is the soliton diameter. At D=0.1 m we obtain S=2 m. Needless to say it is a very rough estimation determined the order of magnitude only. But they show that a flying airplane attracts all BLs located directly ahead, as well as to the right, left, at the top and bottom in the region of inhomogeneous atmosphere produced by the airplane. Achieving the airplane, BL stops in the region where the air density is maximal and moves with this region together and, therefore, with the airplane. No airplane maneuvers can separate BL and the airplane. We ought to note that the additional air pressure Δp near the front edge of the wing is proportional to the square of the airplane speed u, i.e. the region from which solitons are attracted to the airplane increases with an increase in the airplane's speed. Besides, the probability for an airplane to meet any BLs is proportional to the airplane speed. Indeed, if the airplane could do several revolutions around the Earth per one second, it could gather all BLs which were met at its trajectory. Are there any other hypotheses that explain how BLs can pursue an airplane and moves together with it? Such hypotheses are absent.

As is known the air is rarefied at a great height where airplanes fly. Because of this the additional air pressure is generated in an airplane cabin and lounge to provide normal breathing for passengers. Since it is very hard and is not justified economically to keep these rooms hermetic absolutely, there is an air compressor which continuously pumps in the outdoor air. It is natural enough to take the air from that region where the air pressure is maximal that is from the region where BLs are located. The air is pumped by a turbine compressor and there is a tunnel between the lounge and outdoor air at any position of turbine blades. The air density gradient is directed into the lounge. Moving along the gradient, BL penetrates into the lounge. Mechanisms which enable BL to change its shape are considered in [5, 6, 7, 8]. Thus, BLs are specific passengers. They can catch up to a flying airplane and come in without any ticket. One can give a recommendation to airplane designers to prevent

BL penetration. There must be no splits in the region where the additional air pressure is maximal.

The following case which had occurred in the Vologda area in February, 1946 is rather indicative. The second pilot had seen how on the right wing of the plane around a green fire there was a bright white sphere. It was thought that there was a short circuit of an electric lamp, but the flash had not disappeared, as usually happens. The sphere had slowly spread on the frontal edge of the wing and had disappeared under the nose part of the machine. Then a loud crash was heard and black smoke entered into the pilot cabin. Communication had broken off. The commander asks the navigator: "Nick, did you notice the place from which the sphere rolled out?" “In fact it had appeared directly near your legs" the navigator answered: "I have taken the gun for the rocket to check up, what color a charge within it. But I had no time to open it. The blinding white sphere flashed at the very same time. It, as an eye of a devil, peered at me and then had floated to you".

We should note here two circumstances. The sphere moved along a frontal edge of a wing, where pressure of the air is as high as possible. The sphere had got into a cabin through the hole into which a gun for rocket is inserted. This hole is designated to push rockets outside and, therefore, the hole is a tunnel between indoor and outdoor spaces. The air pressure in a cabin of the plane is greater than that outdoors, because air is rarefied at the great height where planes fly. At the same time the air pressure in the cabin of the plane for the comfort of pilots is supported to the pressure close to a normal one. Therefore the sphere, moving in the direction of the gradient of the air density, got into the cabin through the hole in which the gun for rocket was inserted.

A description of collisions of planes with BL damaged or melted off rivets in a forward part of a fuselage are mentioned [9]. It is easy to explain because the rivet edge on a smooth surface of a fuselage breaks a uniform air flow near the fuselage. The excess pressure is created near the rivet, and there is a local maximum of the air density. Having gotten into this maximum, BL evaporates a part of the rivet. This action is explained by the same mechanisms which are responsible for evaporation of long metal cylinders. Evaporation occurs until the edged part of the rivet disappears that leads to a disappearance of the excess pressure created by the edged part of the rivet.

In our opinion, an explanation of mechanisms which enable BL to catch up and accompany airplanes is a strong argument in favor of an optical nature of Ball Lightning. No particles can accompany an airplane. They will be blown out by a stream of running air immediately. On the contrary, it does not matter for light circulating in the space where air pressure is maximal whether motionless or moving air produces the excess pressure, because the speed of moving air is smaller by million times than light speed.

SOLITON BEHAVIOR NEAR METALLIC OBJECTS

Now consider the question why BL is not indifferent to metallic things. There are many reports that BLs are attracted to metallic objects such as wire fences or telephone lines. When attached to metallic objects BLs generally move along those objects. The point is that all metallic objects offer not only good electric conduction but also good heat conduction. Because of 24-hourly temperature changes the temperature of various objects at the earth surface is not identical. On a sunny day the temperature of the upper layer of the earth's surface is greater than that of layers located more deeply. For example, if a metallic pole is driven into the earth, the pole temperature is lowler than that of the surrounding air because the heat which the pole obtains from the air propagates downward to the bottom end of the pole. As a result, in the region near the pole the air density gradient is directed to the pole surface. Appearing in the region, a soliton begins to near the pole. If the air refractive index gradient is not perpendicular to the pole axis then the soliton, approaching pole and heating the air layer between the soliton and pole, begin a motion along the pole axis at some distance from the pole surface. In general, the soliton obeys the first Newton law of uniform movement. Indeed, if the soliton moves in some direction in a homogeneous air, the soliton leaves behind it the air which is warmer than that before it. This is favorable to further movement.

For example evidence number 8 is described, in [3] where two Ball Lightnings appear from stove 5 (figure 11), pass through 2 doors, and along walls approach grounding 6 in a form of a metal wire. A trajectory of Ball Lightnings is shown in figure 11. The BL motion is observed by 4 witnesses 1, 2, 3, 4. Approaching the wire between windows 7, BLs change their shape in an elongate ellipsoid, penetrate though a split between the wire and floor, and disappear.

Penetrating through the chimney in stove 5, solitons move in the direction where the air is cooler. Since the wire for grounding is buried deeply in the ground where the temperature is lower than that in the room, the wire is cooler than the air surrounding it. The wire produces near itself the refractive index gradient directed downwards.

It is known with assurance that BLs can damage wiring, burn fuses, screw plug fuses etc. One can ask if BL is not connected with electricity perfectly, what is responsible for these phenomena. A credible speculation is proposed by Stakhanov [3] who supposes that events occur in the electrified space where the strength of the electrical field is about 3–30 kV/cm. An explosion of BL entails redistribution of the electrical parameters of the space, in particular, the electrical field and voltage in the large enough volume. Such redistributions can be sensed by people as an electrical shock. Besides, as is known, the voltage of an electrical conductor is constant at a steady-state and there are no currents within the conductor. A redistribution of the electrical field forces a redistribution of electrical charges at the conductor surface that is the current in the conductor. This current can be a reason for the above-mentioned troubles.

In some cases interaction of BL with metal subjects gets another characteristic. It occurs when the geometry of a metal object is that which BL can adjoin only to its small part. For example, BL adjoins to an end face of a metal wire. In this case conditions of distribution of heat in metal differ from the case when BL can adjoin with the metal body whose sizes in all three dimensions are comparable. In the first case heat can propagate only in one measurement—along a wire. In the second case, heat can extend in three measurements. Therefore in the first case heating of metal in the place where BL is located is much greater than in the second one.

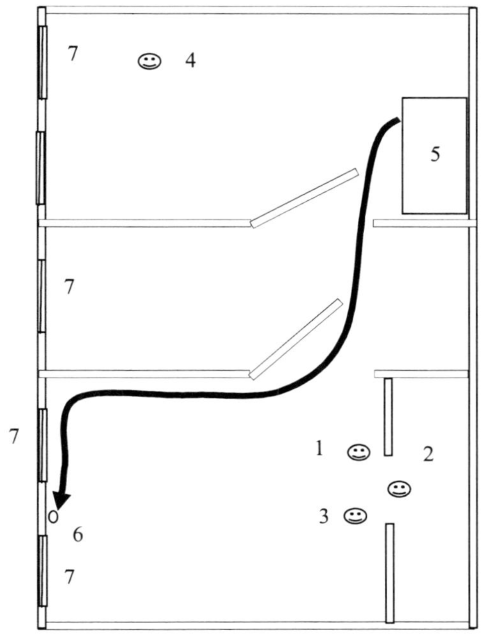

Figure 11. Trajectory of BL from stove 5 through 2 doors, along walls to grounding 6 in a form of a metal wire

If the temperature of metal rises so that evaporation begins, the situation changes radically. The refractive index in the space where evaporation takes place increases due to increasing the refractive index because of the insertion of new portions of matter in this space. Since a soliton tends to the space where the refractive index is maximal, the soliton locates near the place. The soliton evaporates the wire until either its energy exhausts or the wire evaporates completely.

The presented picture has numerous confirmations of eyewitnesses. The witness saw, how after an extremely strong strike of lightning in a telephone column a bright ball of 1 m diameter rolls along a street and after it some more spheres of smaller diameter [10]. It was revealed later that the wire between two columns was torn off and its significant piece disappeared.

In another case a very strong lightning strike had destroyed a copper antenna after it had been installed for studying storm electricity.

The witness observing a thunderstorm had noticed the big fiery sphere, apparently formed owing to vaporization of 65 m wire of 2mm in diameter.

Stakhanov in the case of number 4 describes a case, when BL burned out space of 5 mm diameter and 3 mm depth at an end face of a ramrod attached to a gun. It is marked, that there were no traces of welding. Metal had simply disappeared.

LARGE SOLITON AND FLYING SAUCERS ARE THE SAME

It is very interesting which shape a large soliton of several meters diameter takes. In accordance with our approach, any soliton tends to the space where the refractive index is maximal. What happens if the refractive index changes along height in such a way that its changes are distinctly at the distance equal to the soliton diameter? In this case the soliton equator is located in the layer where the air density is maximal as is shown in figure 12. The top and bottom poles are located in layers where the air refractive indexes are smaller. In this case the air density gradient in the layers is directed to the equator. As a result, there are forces above the equator that tend to draw the soliton top surface to the equator and there are forces below the equator that tend to come close to the bottom surface to the equator. In other words, there are forces which tend to flatten the soliton.

It was shown that inner forces that provide a spherical shape of soliton are small enough to such a degree that the soliton changes its form in inhomogeneous air space. In particular, BL takes the shape of a sausage to penetrate into a room through a hole whose diameter is smaller than the soliton's. The forces which provide the spherical shape of the soliton are in inverse proportion to the square of the soliton diameter. In this case the forces which tend to flatten the soliton increases with increase in its diameter but the forces which provide spherical shape for the soliton decrease. As a result, a change of the soliton shape may be noticeable and the soliton is transformed from a sphere into an ellipsoid of revolution.

If the air density gradient distributions are not identical at the top and bottom poles (typical case), the deformations of the soliton shape are not identical also and the soliton is shaped into a body of revolution resembling a saucer or plate.

Among many known various fantastic properties of flying saucers there is decisive evidence that they can move with high speed and change direction so sharply, that overloads arising from it could not be tolerated by living beings. If flying saucers are in fact solitons, they can move at high speed provided that the refractive index gradient in the air atmosphere is great enough. In the event that an atmosphere is non-stationary, there can be areas in which the direction of the air density gradient changes sharply enough. In such areas solitons sharply change the direction of motion. In accordance with (6) the soliton speed is proportional to the soliton diameter. Therefore it is no surprise that the great solitons can move at a high speed.

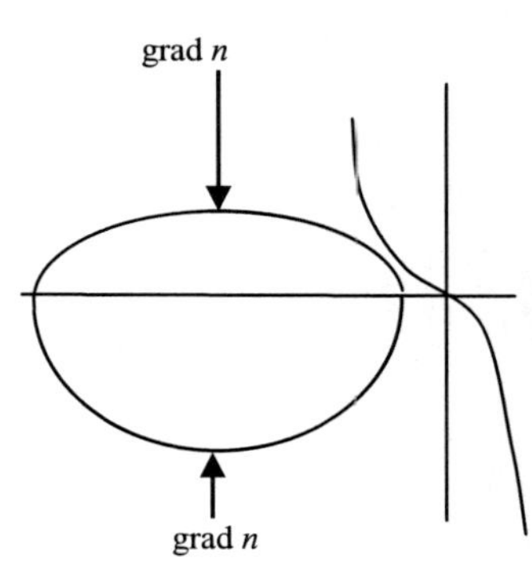

Figure 12. Form of BL of great diameter in a horizontal layer with maximal refractive index. Dependence of grad n jn height is shown at the right

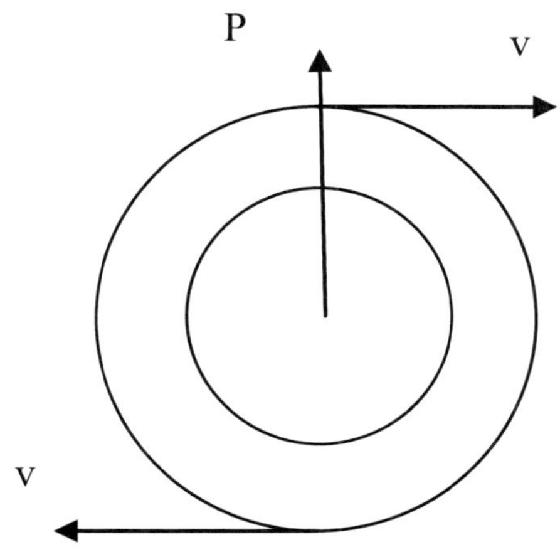

Figure 13. Directions of the air speed and the gradient of the air pressure in vertex

Why Directions of Wind and Soliton Motion Can Be Different

To provide one more solution of abnormal properties connected with BL, we present an explanation why the direction of BL movement can be different from the direction of a wind. The explanation of this BL feature, alongside with an explanation of the mechanism of penetration of BL through glass and the mechanisms allowing BL to catch up and accompany planes, is the third powerful argument in favor of an optical BL nature.

The explanation is simple enough. A soliton moves along the refractive index gradient whose direction can be different from the direction of the air speed. This explanation is complete; however, it is meaningful to give some additional arguments. In figure 13 a particular case of a vertex motion of air is presented. In this case the air velocity is directed along a tangent to a circle. As for the air pressure, it is identical on this circle owing to an axial symmetry of the system. Then the tangential component of the pressure gradient is equal to zero and the gradient is directed along radiuses. When approaching the center, the air pressure decreases. It provides for the occurrence of centripetal forces producing movement of air with centripetal acceleration, directed to the center of the circle. As the air refractive index is proportional to air pressure with other things being equal, the air refractive index gradient is proportional to the air pressure gradient. Thus, the refractive index gradient is perpendicular to the air speed and the soliton in this case moves perpendicularly to the wind.

By the way, if an air moves in some pipe or in some crack, the direction of the air pressure gradient is opposite to the air velocity. In this case, unlike movement of all known objects along the air velocity, the soliton moves against the air speed. Indeed, the air moves from the space with the high pressure to the space with the lower pressure. On the contrary, the soliton moves from the space with the low pressure to the space with the high pressure.

Why the Soliton Seems Cold

One more additional argument in favor of the optical BL nature is an explanation of the following seeming contradiction. The majority of physicists if not all are perplexed why BL radiates white light whose spectrum corresponds to the temperature of the body heated at some thousand degrees, and at the same time BL is rather cold as it does not burn objects near it. There is a simple explanation of this paradox. The light which radiates BL is not produced by exited atoms as it takes place at radiation of light by any heated body. The light radiated by BL is the light which appears as a result of scattering of intensive light, circulating in the BL shell. Really, white light has been produced by excited atoms of air at the moment of gas discharge at a strike of conventional linear lightning when BL was generated. This light has been caught in a light bubble and was circulating within for a rather long time. BL can already be far from the place of its appearance, and the air by this time can already have cooled down. Thus, the nature of the light radiated by BL differs from the light radiated by the heated body. BL has caught and has preserved the light radiation which has arisen at a strike of a usual linear lightning, and gradually radiates it in a form of scattering light. The intensity of the scattering light is small and because of this the light is cold.

Why Soliton Hisses and Causes Radio Interferences

Being at a steady-state, the soliton shape is unchanged and there is no noise from the soliton. Considering the soliton as an elastic shell, we need to recognize that there is a majority of eigen frequencies for mechanical oscillations of the shell. These oscillations may be excited by various inhomogeneities which the soliton meets in its way, in particular, particles of dust. If the eigen frequencies are distributed along the sound spectrum uniformly, a hissing sound may be heard. A similar sound may be heard in a radio dynamic if the gain of an amplifier is greater than the nominal one.

As far as radio interference is concerned, although an influence of BL on a radio receiver is mentioned by many eyewitnesses [11], it does not mean that BL generates radio waves as many scientists believe. Cracks from BL were heard in a telephone tube, too [3], which as is known is not capable of receiving radio waves. The following explanation may be proposed. BL can have electrical charges. Having oscillated in its form at sound frequencies, BL causes a redistribution of the voltage in the space surrounding it. Such redistribution is responsible also for electrical effects at BL explosion. The same effect causes cracks in a radio receiver and telephone tube. One can recall cracks in a radio receiver when electrical lamps are turned on or turned off.

Why the Soliton May Be Variously Colored

There are many reports in which BL colors are various. Along with this there are many various explanations of these observations. For example Turner [1] believed that much of the variability in the color of ball lightning results from traces of contaminants. These can presumably be present as a result of the formation of a ball around a solid nucleus or contaminants with dust. Certainly, greenish balls have been seen to detach themselves from what one would expect to have been copper wires. He attributed the blue halos observed around some (presumably clean) balls to common charge neutralization reaction which just happens to be endothermic on some occasions. He also suggested that the very common red or yellow color might result from organic contaminants.

We give our own explanation why BL color can be various beginning reddish and ending bluish whereas the color of usual linear lightning is white. Most simply, an answer to this question is possible. The temperature of area in which BL appearance takes place can vary greatly depending on conditions of gas discharge. As is known, the maximum of a spectrum moves in a short-wave part of a spectrum at an increase in the temperature. As BL represents a combination of the compressed gas and intensive electromagnetic radiation, and the maximum of a spectrum of electromagnetic radiation can change over a wide range also, BL color, which is defined by a maximum of this spectrum, can be too over a wide range. On this simple explanation various subtleties can be imposed. For example, BL appears as a result of numerous merges of tiny BLs of small diameter whose radiating losses depend on wavelength of the light circulating in them. Most quickly such BL loses the radiation in a long-wave part of a spectrum. As a result, the average wavelength of the spectrum moves in a short-wave part. In this case BL, formed because of the merging of numerous tiny LBs, has a violet shade. Such LBs were observed in experiments [10].

Soliton Disappearance

As for the process of a soliton disappearance, it is fast enough. The soliton disappears when it becomes instable because of continuous radiation losses (although soliton stability is conventional because there is no steady-state due to continuous radiation losses). As processes with participation of light have rather small time constants, transient processes in the soliton are very quick. The air pressure gradually decreases with gradual decreasing of the light intensity. This entails increasing radiation losses and, therefore, further decreasing the light intensity and so on. In the long run the light ceases to be held by the compressed air and is radiated in all sides. The compressed air is no longer held by the light and begins to expand. Time constant of this process is determined by the sound speed in air which is smaller by six orders of magnitude than that of light. A small popping or great explosion may be heard. It's often reported that the sound of the explosions resembles that of a pistol being shot. This testifies that the amount of the compressed air is relatively small and the air compression is great enough.

If the soliton disappears silently it testifies that the soliton shell consists of moderately compressed gases whose refractive index is greater than that of the air rather than of strongly compressed air.

Protection against Ball Lights

Taking into account the above consideration, we can draw certain recommendations on how to avoid damage from the BL. The most radical way is to decrease the air density in the room. In this case the outdoor air is drawn into the room but the BLs, contrary to expectations, are pushed off the room. An increase in the room temperature results in the same effect. If the BL appears to be in a room, it is not recommended to wave your hands or, for example, to drive it away with a besom because a region with excess air pressure is generated between the BL and besom. The BL will then approach the besom. It is not recommended to blow at the BL because the BL will come near the region with the excess pressure; this region is the source of the wind. One can move a plain non-transparent object closer to the BL, for example, a great book with a dark cover. After heating the air layer between the BL and book, the book will push off the BL. We propose to readers to explain the behavior of BL if a bullet is shot into it. In our turn we will try to prove that BL was a possible reason for the Chernobyl tragedy.

Analysis of the reasons for this tragedy has been carried out by many researchers. The official version declares that the tragedy grew out of mistakes by the personnel aggravated by a fatal coincidence of circumstances and some constructional deficiencies of the reactor. Undoubtedly, such mistakes by the personnel and constructional deficiencies took place. But was it the definite reason for the tragedy? It is impossible to prove or disprove that. There is a simple explanation. There was a road and a transport incident on a highway at which a car appeared in a ditch. Investigating the reasons for the incident, a car inspection established that the driver had considerably exceeded the speed limit, and the car was technically out of order: bad brakes, inadmissible defects in a steering management, etc. All of this could have led to the incident. However more careful research has shown that directly ahead of the incident a truck came too close to the car on a counter strip, and the driver needed to move to a ditch to avoid a collision. There is a question. Possibly, were there any unknown reasons for a truck to be moving on a counter strip? It appears that such a reason could be, something apparently harmless at first sight, the electro technical tests of the electro generator that began just one minute prior to the tragedy. As a result of these tests, BL could have appeared which penetrated into the nuclear reactor zone and caused it to become out of order.

It is well-known that quite often Ball Lightnings arise in accumulator submarines when switching electro motors. Such Ball Lightnings fly on cabins and sometimes burn racks and burn partitions. Numerous certificates of occurrence of fiery spheres are described in the monograph of Barry [12], devoted to Ball Lightnings. For example, Nilson, a Norwegian expert in hydro energy, has shown that reddish shining matter was separated and remained visible in the air within several seconds at a quick close of a direct current generator. A photo of the shining sphere which had separated from the shining matter is presented in the book. There is evidence that a Ball Lightning about 20 cm in diameter burned a hole of 7 cm diameter in a lateral wall of a factory pipe ([3], case 34), and also numerous evidence how Ball Lightnings appearing in interiors of planes burn their covering. Holes of the correct round form burned into metal targets (aluminum, titan, tungsten) can be seen in the photos in figure 22 in book [14].

The eyewitness of the tragedy, physicist Urutskoev, asserts that a burned out hole of a regular circular form was in the wall of the mine of the nuclear reactor. Various substances which are prone to erosion at the gas discharge can be used. For example, in experiments on production of artificial Ball Lightnings, the wax preliminary put on electrodes between which took place the discharge was used [10]. Metals, various polymers, cotton wool, shavings of a tree were used in other experiments [15]. Apparently, the transformer oil manifested itself as such substance at production of the Ball Lightning in the machine hall. According to the chief of the team who produced the test, there was a fire at the bottom floor and transformer oil was burned [16]. After appearance of a Ball Lightning at erosive gas discharge, events could occur as follows. The Ball Lightning could have gotten through a hole burned into a steam line and, moving along the refractive index gradient, could have gotten into a place where water is converted into a vapor and, hence, "erosion" of water takes place. Density the pair and, hence, the refractive index in this place are maximal. Having reached this area and having destroyed the heat-producing element, the Ball Lightning could have gotten into the mine. After then BLs would be drawn to a metal leaf of the mine, in which it burned out a hole of the regular circular form. After disappearance of the Ball Lightning, vapors of transformer oil compressed in the BL shell would have remained. They could have spread out and gradually condensed on the walls of the mine. As Urutskoev noted, the walls of the mine had been covered by an oil emulsion.

CONCLUSION

Believing a priori, that light bubbles are an objective reality, we have shown that in accordance with well-known laws of physics behaviour of light bubbles in the conventional air atmosphere is completely identical to a mysterious and paradoxical behaviour of Ball Lightnings. Too many riddles connected with Ball Lightnings can be explained under an assumption, that BLs are light bubbles.

It is generally accepted that BL has properties whose combination is absent in any known object. Indeed one can imagine the object which moves against wind. This is a bird. One can imagine the object which penetrates window panes. This is light. One can imagine the object which radiates light. It is a firefly. One can imagine the object which can catch up to a flying airplane. This is other airplane. But it is impossible to imagine the object which has all these properties simultaneously. It must be something radically new because all attempts in many centuries to explain BL nature on the basis of known knowledge ended in failure. None of the 200 existing BL theories can explain a combination of BL properties. A perfectly new unknown up to now object which has all BL properties is required. We have shown in this section that our light bubble has simultaneously all listed properties. It is improbable that it is a simple fortuitous accident.

From the presented consideration, it is necessary to draw the conclusion that light bubbles are not caused by the mind playing games, but are real objects. Till now light bubbles were studied neither theoretically nor experimentally. No one suspected an opportunity for their existence. At this point it would be possible to finish activity on an explanation of the nature of Ball Lightnings and to expect a general recognition of the presented hypothesis.

Unfortunately, new theories win a general recognition not as promptly as we would like. For example, about a century was required for Copernicus's theory that the earth rotates around the sun to win a general recognition. Having no opportunity to wait that long, we undertook further steps in search of new additional arguments in favor of light bubble existence. Properties of objects for which there is still no generally recognized name and which have been produced in laboratories by scientists in the last two centuries at attempts to produce artificial BLs have been analyzed and investigated. This activity resulted not only in additional evidence to the validity of the accepted approach but also disclosed new physical effects deduced from analysis of experimental data obtained from investigation of the objects. These problems are considered in the next section.

References

[1] Turner, DJ. *Phys. Rep.,* 1998, 293, 1–60.
[2] Smirnov, BM. *Phys. Rep.,* 1987, 152, 177.
[3] *Stakhanov, IP. Physical Nature of Ball Lightning*; Atomizdat: Moscow, 1996.
[4] Grigoriev, AI; Grigorieva, ID; Shiryeva, SO. *J. Scientific Explorations,* 1992, 261–279.
[5] Torchigin, VP; Torchigin, AV. *Chemistry and life,* 2003, № 1, 12–15.
[6] Torchigin, VP. *Docl. Phys.,* 2003, 48, no. 3, 108–111.
[7] *Torchigin, VP. Lomonosov,* 2003 no.2, 86-90.
[8] Torchigin, VP; Torchigin, AV. *Physica Scripta,* 2003, 68, 388–393.
[9] Singer, S. *The Nature of Ball Lightning;* Plenum Press: NY, 1971.
[10] *Dmitriev, MT. Priroda,* 1967, 6, 98.
[11] Klimov, AI; Mishin, GI. *Pis'ma Zh. Tekh. Fiz.*, 1993. 18 (13), 19.
[12] Barry, JD. *Ball Lightning and Bead Lightning;* Plenum Press: NY, 1980.
[13] Torchigin, VP; Torchigin, AV. *Europhysics Journal,* 2005, D 36, 319–327.
[14] *Avramenko, RF; Nikolaeve, VI; Poskacheva, L. P* In book *Ball lightning in laboratory., Avramenko, RF; Ed;* Himiya, Moscow, 1994, 7-56 (in Russian).
[15] *Emelin, SE; et al. Zh. Tekh. Fiz.,* 1997, 67, № 3, 19–28.
[16] *Gorbachev, BI.* In book *Chernobyl Problems*, issue 10, part 1, Chernobyl, 2002.

Space Spherical Solitons and Natural Radioactive Decay

Abstract

An attempt to extrapolate a notion of light bubble for nuclear processes is undertaken. Indeed, there is a natural radioactive decay at which nuclei decays abruptly in the same way as light bubbles do. An existence of light bubbles is explained by wave phenomena. In accordance with quantum mechanics an existence of nuclear objects is explained by the wave phenomenon too. From the similarity of nuclei and light bubbles follows that because of inevitable radiation losses inherent to waves of whispering gallery type the energy stored in a nucleus should gradually increase in time. It turns out that this does not contradict the generally accepted fact that the nucleus energy is unchanged in time. It is shown that gradual decreasing of total energy of a system consisting of many identical nuclei manifests itself in abrupt decrease of number of nuclei in the system. Since nuclei with the smallest energy disappear from the system, the average energy of remaining nuclei is constant in time. Features of interaction of a system of nonlinear identical oscillators are considered. Extrapolation of obtained results on a system of identical nuclei enabled us to conclude that quickly prepared long-living gas of densely packed molecules is favorable for an increase of the speed of radioactive decay. Numerous experiments confirming this conclusion are considered.

As is known, introduction of oscillators into consideration has allowed Planck, Einstein, Debye and other physicists to develop an absolutely new approach to an explanation of many physical phenomena. As a classical oscillator was imagined as a mechanical oscillatory system, further research in this direction has led to the appearance of quantum mechanics although it deals with waves in a form of wave functions. Studying of BL nature has allowed us to introduce into consideration other types of oscillators, in particular, to study oscillators where whispering gallery light waves are involved. The lifetime of such wave is finite because of inevitable radiating losses inherent to waves of such type. The further comparison between the oscillators and BL has led to consideration of the self-confined waves of whispering gallery type. In modern terms such formations are optical spherical spatial solitons (OSSS). OSSS represents the resonator for a wave of whispering gallery type formed by this wave in a nonlinear optical medium. Unlike the usual optical resonators for waves of whispering gallery type [1] in which the light wave of any amplitude can circulate, OSSS can exist only provided that the amplitude of a circulating wave surpasses some threshold. When because of inevitable radiating losses the amplitude falls below this threshold, OSSS becomes instable and disappears in the same way, as BL does.

The same occurs to nuclei of chemical elements. They sharply cease their existence with breaking up and forming other nuclei. This phenomenon has been discovered in the beginning of the 20th century and is called natural radioactive decay. Since this phenomenon reminds us of the existence and disappearance of BL, it is interesting to investigate this similarity in more detail. It is especially interesting, as BL and a nucleus are presented by waves. Similarity in the nature of tunnels for light waves in optics and for wave functions in quantum mechanics is so great, that optical analogies are often used for an explanation of tunnel effects in nuclear physics [2]. By the way, a wave function of a nucleus can be a whispering gallery type too. Certainly, this wave function also satisfies to a wave equation. In this case similarity becomes complete, and all the conclusions obtained for self-confined waveguides for waves of whispering gallery type [3, 4, 5] which are a basis of BL theory, can be applied to wave functions too.

A hierarchy of various forms of spherical space solitons is presented in figure 14. Till now we considered the left part of the figure where IOSSS is an abstract notion or theoretical model and BL or AO are its real representatives in nature. The right part presents a particular case of SSS where self-confined whispering gallery waves are wave function type. Nuclei of various chemical elements are its real representatives in nature. It is interesting to consider consequences of such an assumption. Like IOSSS where the intensity of light waves of whispering gallery type gradually decreases in time because of inevitable radiating losses inherent to whispering gallery waves, we need to accept that wave function of SSS decreases gradually in time too. Moreover, we should accept that the square of the module of a wave function characterizes the density of energy, rather than the density of probability as it is usually considered.

Seemingly, these assumptions contradict the experimental facts according to which energy of any nucleus is constant in time. Besides, as follows from these assumptions, all nuclei should be stable during all their life, and then they should break up simultaneously when their energy becomes below a certain threshold. It has appeared that these fears are deprived bases. Really, considering nuclei as specific spherical spatial solitons, we should agree that nuclei interact one with another and they exchange their energy. From here follows, that energies of nuclei are distributed in some narrow interval. When energy of some nucleus becomes below a certain threshold, it breaks up and disappears from the set of identical interacting nuclei. Disappearance of the nucleus whose energy is smaller than the average energy in the set entails an increase in average energy of the remaining nuclei. This shift of average energy is compensated by a gradual continuous and monotonous reduction of energy in time of all remaining nuclei because of radiating losses. As a result, average energy of nuclei remains constant. External characteristics of radioactive decay completely coincide with experimentally observable ones.

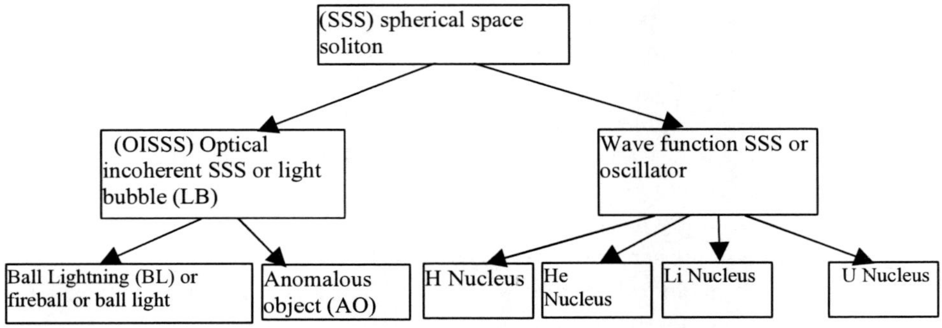

Figure 14. Hierarchy of various forms of spherical space soliton and their designations

This conclusion is valid in the case where a system of nuclei is in a steady-state. If somehow the steady state is violated, at this time the speed of the radioactive decay can be changed.

INTERACTION OF A SET OF IDENTICAL OSCILLATORS IN A FORM OF THE SELF-CONFINED WAVE FUNCTION

Such oscillators are characterized by a feature that the energy stored in an oscillator decreases gradually in time because of radiating losses. As a result, all oscillators are in the common field produced by them. As is known, radiating losses of an optical oscillator in which a light wave of the whispering gallery type is circulating are defined by parameter $N = 2\pi r_0/\lambda$, where r_0 is radius of sphere, λ is wavelength of light wave. As shown in [6], the quality index determined by radiating losses for a glass ball with $N = 2000$, is equal $Q = 10^{400}$. Energy of oscillations in the resonator decreases as follows

$$E = E_0\exp(-2\omega t/Q) = E_0\exp(-\gamma t), \quad (8)$$

where $\gamma = 2\omega/Q$. Assuming, that for a light wave $\omega = 10^{15}$ c^{-1} and lifetime of light is defined only by radiating losses, we obtain an unimaginably large value equal to 10^{378} years.

Consider now interaction between a set of chaotically moving identical nonlinear oscillators in which the stored energy gradually decreases in time. Similar to molecules of gas in a closed vessel, oscillators can casually collide one with other. Thus every oscillator can either acquire or lose any energy at collision. Finally, a normal distribution of energies of oscillators comes (figure 15) at which a probability p of an oscillator to have energy in a range from E to $E + \delta E$ is equal to $p(E)\delta E$, where

$$p(E) = \sigma^{-1}\pi^{-1/2}\exp[-(E - E_0)^2/\sigma^2], \tag{9}$$

$\sigma = F(k)$, $F(k)$ is some increasing function.

As follows from (9), $p(E_{th}) > 0$, where E_{th} is the threshold at which the oscillator should disappear. Since the number of nonlinear oscillators in the system decreases, the distribution (9) is pseudo steady-state. Disappearance of an oscillator with minimal energy E_{th} is accompanied by an increase in average energy of remaining oscillators $<E>$ by $(E_0 - E_{th})/N$, where $N >> 1$ is the number of oscillators in the system. Indeed, the average energy of oscillators can be presented as follows:

$$\langle E\rangle = N^{-1}\sum_{i=1}^{N} E_i = N^{-1}\sum_{i=1}^{N}(E_0 + \Delta E_i) = E_0 + N^{-1}\sum_{i=1}^{N}\Delta E_i \tag{10}$$

where E_i is the energy of i-th oscillator, $\Delta E_i = E_i - E_0$ and $\sum_{i=1}^{N}\Delta E_i = 0$

Assume that an oscillator with number N disappears. In this case its energy is equal to E_{th}, and the term in (10) for which $\Delta E_N = E_{th} - E_0 < 0$ disappears too.

Then $\sum_{i=1}^{N-1}\Delta E_i = (E_0 - E_{th}) > 0$

and

$$\langle E\rangle = E_0 + (E_0 - E_{th})(N-1)^{-1} \tag{11}$$

If the number of disappearance per second is λN, then the total energy increases by $\lambda(E_0 - E_{th})$. This increase is compensated by a decrease in total energy because of inevitable radiating losses in each oscillator in accordance with (8). Then

$$dE/dt = -\gamma E(t) \tag{12}$$

The decrease is accompanied by exchanges of portions of energy at collisions of oscillators. Such exchange promotes the idea that distribution of energy between oscillators tends to a normal one. This tendency is violated by disappearance oscillators, whose energy is smaller than threshold E_{th}. Thus, the gradual decrease in energy in every oscillator is transformed to a sequence of discrete decreases of the system energy due to disappearance oscillators. Speed of the constant decrease in the total energy is defined (12) and is in agreement with the speed of decreasing number of oscillators that is determined as follows

$$dN/dt = -\lambda N(t), \tag{13}$$

where N is total number of oscillators in the system, λ is a probability of disappearance of oscillators per one second. Since the total energy in the system is proportional to the number of oscillators in it, then $dN/N = dE/E$, and from (12), (13) we have

$$\lambda = \gamma. \tag{14}$$

Note that this relation depends on neither internal parameters of oscillators, nor on the parameters describing their collisions, such as strength of interaction characterized by dimensionless coupling index k, duration of collision τ_R, frequency of collisions. This conclusion is valid for any system in steady state when average speed of gradual increase in the total energy is equal to average speed of increase because of disappearances of oscillators. Let show that this steady state is stable. Let the curve 1 in figure 15 represents normal distribution of the density of probability for oscillator to have energy E. The area under this curve is equal 1. The curve has a maximum at $E = E_0$, and its width depends on the coupling index between oscillators at their collision. If energy of the some oscillator becomes equal E_{th}, oscillator disappears. The smaller difference between E_0 and E_{th}, the greater is probability of its disappearance.

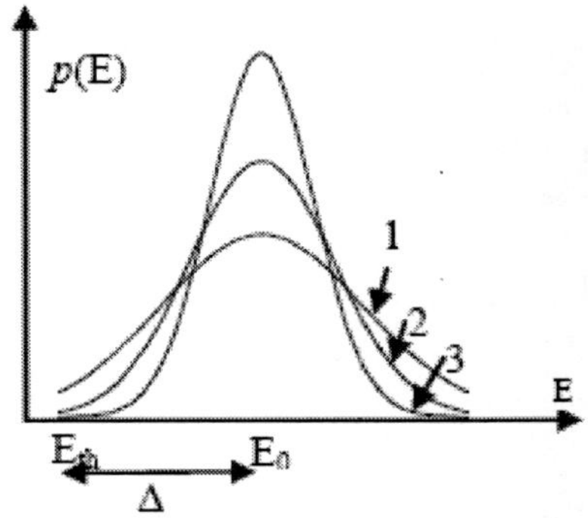

Figure 15. Dependence of density of probability p of appearance of oscillators on their energy E

Suppose that maximum of curve 1 takes place at $E = E_1$ where $E_1 < E_0$, that is the curve 1 is shifted to the left. In this case, as follows from figure 15, the speed of disappearance of oscillators raises as compared with the speed at steady state when $E = E_0$. As has been shown, a disappearance of each oscillator is accompanied by an increase in average energy remained oscillators by $\Delta E/N$, where $\Delta E = E_1 - E_{th}$. The greater the speed of disappearances is the greater speed of an increase in average energy. As a result, E_1 increases. This leads to shift of curve 1 to the right, and the speed of disappearance oscillators with energy E_{th} decreases. Such process proceeds until a steady state will be achieved and condition (14) becomes valid. The case when $E_1 > E_0$ can be considered similarly.

Ought to note this conclusion is valid for any form of curve 1. For example, the same reasoning could be applied for curves 2 and 3. It means that the conclusion about validity of condition (14) does not depend on the coupling index. Notice, that number of oscillators which disappear when their energy becomes equal E_{th}, does not depend not only on probability $p(E_{th})$, but also on frequency of collisions. Probability $p(E_0)$ in comparable with $p(E_{th})$ in figure 15 for the sake of illustrative reasons only. Actually $p(E_0)$ is greater than $p(E_{th})$ by many orders of magnitude.

Under assumption that energy of an isolated oscillator is unchanged in time we receive $\gamma = 0$, and, therefore, from (14) $\lambda = 0$. In this case speed of disappearance of oscillators decreases in time and tends to zero at approach to steady state. This contradicts to experimental data.

Notice that the presented considerations are not valid for transient processes in the system. Consider several examples for the sake of illustration. Assume, that the coupling index becomes equal to zero at $t = 0$. In this case interaction between oscillators ceases and their energies tend to E_{th} independently. The total energy of the system gradually decreases and maximum of curve 1 in fig. 15 moves in a direction to E_{th}. In this case the probability of disappearance of oscillators rises with time. Energy oscillators which will disappear at time instant t, is equal $E = E_{th}\exp(\gamma t)$. Taking into account (9), the probability of disappearance of oscillators can be presented as follows:

$$p(t) \sim \exp\{-\frac{[E_{th}\exp(\gamma t) - E_0]^2}{\sigma^2}\}$$

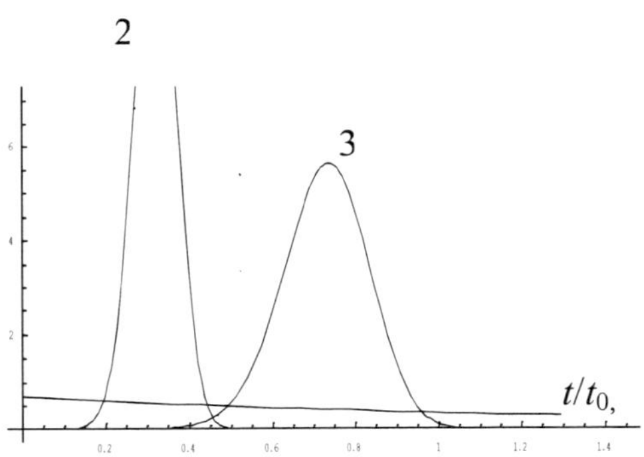

Figure 16. Dependence of radioactive decay on time. 1 – natural decay; 2 – after disappearance of week interaction between oscillators; 3 – after disappearance of strong interaction between oscillators

$$\frac{dN}{dt}(a.u.)$$

Curve 1 in figure 16 shows dependence dN/dt on t in steady state when $\gamma = \lambda$, curve 2 shows the same dependence when the coupling index becomes equal to zero at $t = 0$. It is supposed, that $\sigma/E_{th} = 0.05$ and $Et_h/E_0 = 0.8$. As is seen from figure 16, at $t = 0$, dN/dt for curve 2 is much smaller, than that for curve 1. However at $t \approx 0.3t_{0.5}$ dN/dt for curve 2 surpasses dN/dt for curve 1 approximately by 30 times, where $t_{0,5}$ is the time period in which number of oscillators decreases by two times. Dependence dN/dt on t at $\sigma/E_{th} = 0.1$ and $Et_h/E_0 = 0.6$ is presented by curve 3 in figure 16. In this case maximum of disappearance of oscillators comes later at $t \approx 0.75\ t_{0.5}$.

Thus, if the curves similar to dependences 2 and 3 could be obtained experimentally, σ and E_{th}/E_0 could be calculated, and average value of coupling index k could be estimated.

THE EXPERIMENTAL DATA ABOUT A CHANGE OF SPEED OF RADIOACTIVE DECAY WITH A CHANGE OF EXTERNAL CONDITIONS

According to existing representations "λ is one of the major characteristics of an atom's nucleus: λ does not depend on physical and chemical conditions, such as temperature, pressure, concentration, a chemical compound, from age of a radioactive nucleus" [7]. In our designations it concerns a constant γ, which is determined by internal properties of a nucleus and can not be changed by external forces. In the encyclopedia it concerns a constant λ which can be determined experimentally. According to (14) $\lambda = \gamma$ at any steady state. However these constants can differ during transient processes.

There are enough both old and new works as theoretical so it's the experimental character that points out an opportunity for changing radioactive decay speed. For example, recently it has been shown, that the greatest stability in isobar series, i.e. among elements with identical nuclear weight, has an atom with minimal energy. Not only the very energy of the nucleus is taken into account but also the addition in a form of energy of the electron environment of the nucleus. This energy is equal to 13.6 eV for the hydrogen atom and to several thousand eVs for atoms with $Z > 20$, where Z is number of protons in an atom. This addition is smaller by 3–4 orders of magnitude than the energy of a nucleus equal approximately 931.5*A MeV, where A is atom weight of a nucleus.

Modern accuracy of measurement of atom weights is equal to about 1,000 eV. Thus, now energy of the atom can be measured in view of energy of an electronic environment. The analysis of accessible experimental data [8] has shown that a stable atom takes place only at a minimum of its energy. Such conditions as "the minimum of weight of a nucleus" or "a maximum of energy of connection" appear insolvent [9].

This conclusion can be explained in a natural way on the basis of the presented approach. Really, a continuous slow decrease in the energy stored in an atom leads to passage into a new steady state with smaller energy. Such condition is available for all atoms in isobar series, except for atoms with minimal energy. For this atom decrease in the energy should be immeasurably greater that it could pass in the steady state related with another isobar series. Time required for a decrease in the energy in this case should be immeasurably greater. Differently, time of half-decay of such atom should be immeasurably greater, that is, it should be practically stable.

There are several important consequences from the presented consideration. First, an atom and its electronic environment represent a single whole entity, and it is necessary to consider their general energy. Secondly, difference $E_0 - E_{th}$ is rather small as compared with E_0. In the third, full ionization of the atom reduces this difference considerably, as energy of ionization is comparable with the difference between energies of atoms in common isobar series. As a consequence, full ionization of the atom should change its half-life period noticeably. Experiments [10], (Bosch, 1996) confirm this conclusion. It is shown, that the half-life period completely ionized atom ^{187}Re decreases by 10^9 times as compared with a neutral atom (the half-life period decreases from 4.3 10^{10} years to 33 years).

Influence of an electric field of an atom on the probability of beta decay is considered in detail in surveys [11], [12] where experimental data are presented also. Influence of changes of an electronic environment of an atom on beta decay of a nucleus tritium is considered in detail in works [13], [14] where convincing experimental data are presented also.

It is shown [15] that radiation of uranium after electrolysis increased by 2 times as compared with initial material U_3O_8. As a concentration of uranium after electrolysis has appeared at 30 % higher than in the initial material, the normalized emission of α-particles in uranium after electrolysis became 30 % greater than in a initial material.

The presented approach, at which the nucleus is considered as a spherical self-confined soliton in which the stored energy constantly decreases because of radiating losses, enables one to connect a natural radio-activity with gradual decreasing energy of a nucleus. Randomness of natural radioactive decay can be explained not by some unknown processes within a nucleus but random processes outside a nucleus provided the stored energy in a nucleus gradually decreases. If one accepts that energy of a nucleus is constant, the speed of radioactive decay should gradually decrease at the accepted approach. The fact that external conditions can influence the speed of radioactive decay confirms the possibility of the existence of the presented approach.

Theoretical Preconditions for Increasing Speed of Natural Radioactive Decay

A natural question arises. Whether the speed of radioactive decay can be changed to such a degree that energy that is exerted at any nuclear conversion can be used in alternative sources of energy? As has been shown above, the inner energy within a nucleus decreases gradually in time as well as changes abruptly at collision of nuclei. When the inner energy decreases down to the threshold at which the nucleus becomes instable, the nucleus breaks up and some excess energy liberates from the nuclei. Unfortunately the intensity of this process is extremely small for all natural chemical elements. By the way, such process is used in Plutonium power suppliers located on satellites. Their power is about 3 kW and duration of operation is about 80 years. But they are very expensive devices because a very expensive substance is used as a fuel. Usual natural matter widely available throughout nature is desirable to use as a fuel.

The simplest way to increase the speed of a radioactive decay is to increase coupling index k which characterizes strength of interaction. It turns out that it is insufficient. Unlike molecules of gas which are considered as elastic balls at their interaction at collisions and their energy is proportional to the square of their velocity, interaction of oscillators is perfectly different. Firstly, the energy of the oscillator is not connected with its velocity. In this case motionless adjacent oscillators can interact one with another and duration of their interaction can be as great as is wished. Secondly, an oscillator is an excellent accumulator of energy if the speed of transmission of energy into the oscillator is greater essentially than the speed of a decrease of energy due to radiating loses. Reality shows that this condition is held with great because oscillators exist in many billions years. Thirdly, transmission of energy between oscillators depends on phase relations between oscillations within them. Transmission of energy is absent completely at certain phase relations.

Consider features of interaction of a system of oscillators to clarify a possibility of using these features to accelerate their disappearance. Initially the simplest case of interaction of two coupled oscillators is considered.

Interaction of two Interconnected Oscillators

A theory of interconnected oscillators is well developed and is used in many absolutely different applications. We present its results used in the further consideration. In the elementary case the following system linear differential equations describe change of complex amplitudes a_1 and a_2 in identical oscillators 1 and 2, respectively, with identical eigen frequencies ω [16]:

$$da_1/dt = j\omega\, a_1 + jk\omega\, a_2$$
$$da_2/dt = j\omega\, a_2 + jk\omega\, a_1. \qquad (15)$$

Here dimensionless coupling index $k = j\kappa/\omega$, where κ is the coupling index used in [16]. The system (15) has two eigen frequencies

$$\omega_1 = (1 - k)\omega \qquad (16)$$

and

$$\omega_2 = (1 + k)\omega, \qquad (17)$$

which correspond symmetrical $\{a_1 = \exp(j\omega_1 t),\ a_2 = \exp(j\omega_1 t)\ \}$ and anti symmetrical $\{a_1 = \exp(j\omega_2 t),\ a_2 = -\exp(j\omega_2 t)\ \}$ eigen solutions. Any linear combination of these eigen solutions is also a solution of the system (15). Any solution describes beatings which amplitude and phase depend on initial conditions at $t = 0$.

For example, if $a_1(0) = 1$, $a_2(0) = 0$, then $a_1(t) = \cos(\Omega\, t)\exp(j\omega t)$, $a_2(t) = \sin(\Omega\, t)\exp(j\omega t)$, where

$$\Omega = k\omega \qquad (18)$$

is a frequency of beatings which does not depend on the initial conditions. Since the energy stored in an oscillator is proportional to a square of amplitude of oscillations, then

$$E_1(t) \sim \cos^2(\Omega\, t),\ E_2 \sim \sin^2(\Omega t), \qquad (19)$$

where $E_1(t)$, $E_2(t)$ are energies stored in oscillators 1 and 2, respectively. It easy to see that $E_1(t) + E_2(t) = \text{const}$. Dependences of energies of two interconnected oscillators on time at optimal phase relations are shown in figure 17.

A transfer of energy between two identical nuclei depends on phase relations between oscillations in them. A transfer of energy is absent only in the eigen solutions in which phases are identical in the symmetric own solutions or differ by π; for antisymmetric one. Speed of transfer of energy is maximal, if amplitudes of oscillations are identical, and their phases differ by $\pi/2$. The first condition is valid for nuclei.

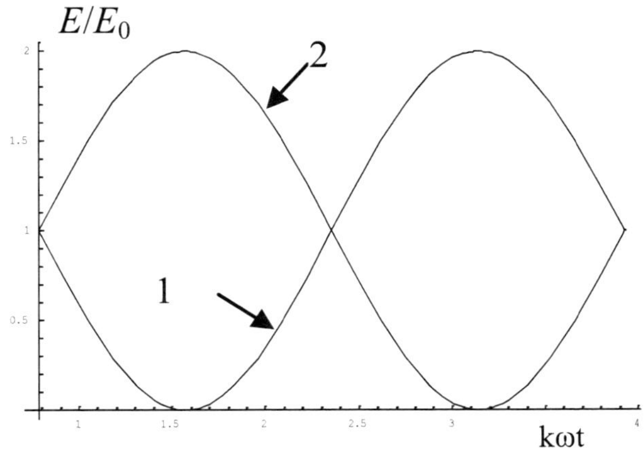

Figure 17. Dependences of energies of two interconnected oscillators on time at optimal phase relations: Curves 1 and 2 show dependencies for the first and second oscillators, respectively

The second condition can appear valid accidentally. In general, transfer of energy takes place at any phase difference φ. However the amplitude of beatings can change from zero (when the difference of phases is equal 0 or π) up to maximal. Transfer of energy takes place at any coupling index k but the speed of transfer is proportional to k. It is possible to conclude, that appreciable interaction between oscillators is possible at any distance between them provided that duration of interaction is comparable with the beating period.

$$\tau_b \cong 2\pi/\Omega = 2\pi/(k\omega) = \tau_n/k. \tag{20}$$

As to coupling index k that it is known reliable that k is extremely small. As shown in [16], appreciable interaction between two oscillators with own frequencies ω_a and ω_b, takes place, if the following condition $(\omega_a - \omega_b)/\omega_a < k$ is valid. In this case the difference between own frequencies of any two unequal nuclei is so great, that this condition is not valid. Thus, a noticeable interaction can be only between identical nuclei.

It is necessary to take into account that the nuclei considered as oscillators are a typical "nonlinear oscillating system". It means that properties of nuclei depend on amplitude of oscillations. In our example, energy of oscillations increases in one nucleus by 12 % and decreases in other nucleus in time $0.01\tau_b$. The difference in amplitudes can become so great, that the stationary condition of nuclei appear violated and the system from two interconnected nuclei can pass into a new steady state characterized by a new local minimum of the total energy which is smaller than the previous low was. Thus superfluous energy is liberated. For example, there can be a synthesis of two nuclei of deuterium. On the contrary, for heavy nuclei, for example, uranium, there can be a nuclear reaction of division. In any case own frequencies of oscillators after liberation of the excess energy become various and their interaction stops. Thus, any noticeable in action R can be only between identical nuclei.

Ought to note that instable states of a nucleus uses in the nuclear physics for a long time for an explanation of that fact, that a slow neutron can split a nucleus ^{235}U better, than a fast neutron. Unlike a fast neutron, a slow neutron can be absorbed by nucleus; it becomes instable and disintegrated as a result of this instability. We are going to show that instable states can appear not only under action of slow neutrons, but also as a result of long interaction of two identical nuclei.

Interaction of a Set of Oscillators

It is simply enough to imagine a picture of interaction of two nuclei located near each another for a long time. But how such picture can be realized? As is seen from (18), (19), the degree of interaction R at the maximal speed of transfer of energy is determined by the expression:

$$R = \Delta E/E_0 = 2\Omega\tau = 2\omega k\tau, \tag{21}$$

where $\Delta E/E_0 << 1$ is relative change of energy in time τ. We have from (21) that R is maximal when product of $k\tau$ is as much as possible at optimum phase relations. The coupling index k can be increased by decreasing a distance r between oscillators. The most radical way to decrease r is to increase density of interacting oscillators.

Seemingly, liquids or solids are the most suitable candidates for realization of great interaction between oscillators because their nuclei are located to each other closer than in gases. However it appears that in these conditions a majority of interconnected oscillators is in a steady state and the difference of phases of oscillations between adjacent interacting oscillators is equal to zero. In this case there is no interaction between oscillators. Interaction of a set of mutually interconnected oscillators is well enough studied in radio engineering as a set of mutually connected oscillators in the form of LC circuits is used in high-quality radio receivers. Nuclei in liquids or solids considered as interconnected oscillators form a specific three-dimensional transmission line in which any deviation from a stationary condition in any oscillator at once begins to propagate in all directions through a chain of interconnected adjacent oscillators.

The more coupling index between oscillators the greater the speed of propagation of the deviations. In other words, oscillators are synchronous among themselves. Certainly, there are some small deviations from steady state because of thermal motions of atoms, but these deviations cannot lead to a sharp increase in amplitude of some oscillators.

Like a random motion of gas molecules can not entail a sharp increase in gas density, small random phase shifts can not entail significant increase in amplitudes of oscillations. Any increase in amplitude of oscillations entails immediate scattering. Thus, liquids or solids do not permit to accumulate results of interactions because phase relations between oscillators are far from required ones.

Absolutely other situation takes place in a set of moving oscillators presented by atoms of gas at normal conditions. In this case time of collision oscillators is much smaller than time of free movement between two consecutive collisions. Each collision is accompanied by occurrence of a system consisting of two interconnected oscillators. In such system its own frequency can change in a range $(1 - k)\omega < \omega < (1 + k)\omega$ and depends on phase relation between oscillations in the oscillators.

The change in own frequency leads to change of phases of oscillators as compared with a phase of an isolated oscillator with own frequency ω. Change of phases of oscillators can be both positive and negative depending on initial phases between interacting oscillators. The total phase shift appeared as a result of one collision depends on force (k) and time of interaction. The smaller distance between interacting nuclei, the greater phase shift at one collision. Thus, unlike system motionless interacting oscillators where take place certain phase relations between adjacent oscillators, the steady state of a system of chaotically moving and colliding oscillators is characterized by random phases of each oscillator.

It is easy to be convinced, that if phase relations at each collision are random, the probability that favorable collisions considerably prevail above unfavorable ones is negligible small. This conclusion does depend on neither pressure nor temperatures of gas. Because of this it is useless to try to increase result of interaction by means of increase these parameters. Thus, neither solids nor liquids nor gases can provide significant change in energy of oscillators. Seemingly, our attempts to use features of oscillators for significant change of their energy are fruitless. However non steady states have been not analyzed yet.

Effect of Repeated Long Interaction

Assume that gas is compressed so much, that oscillators of gas are densely packed in such degree, that the volume per one oscillator is equal to volume of the oscillator. In such gas an effect which enables to accumulate significantly result of interaction R appears. Unlike gas at normal conditions where each oscillator collides each following time with a new oscillator and a probability of collision in the near future with a former oscillator is extremely small, the behaviour of a oscillator in gas densely packed oscillators is absolutely different. Each oscillator oscillates in an environment of the same adjacent oscillators which do not allow it to move freely. In this case two identical adjacent oscillators can interact many times between themselves and accumulate result of interaction R.

Phases of oscillations in homogeneous gas of densely packed oscillators are random; at least, they are random after sharp compression. In this case the probability that the area where R can be accumulated noticeably is rather great. Really, if oscillator is surrounded by K identical oscillators, the probability that phases of all these oscillators are favorable for transmit of energy to central oscillator is equal to $p = 2^{-K}$. For example, if $K = 6$, $p = 1{,}6\ 10^{-2}$. As the total number of oscillators $N \cong 10^{20}$ cm^{-3}, the number of such areas is rather great.

Generalizing (15) on five oscillators, one of which is located in the point with coordinates ($x = 0, y = 0$), and others are located in points with coordinates: ($x = 0, y = 1$), ($x = 0, y = -1$), ($x = 1, y = 0$), ($x = -1, y = 0$), we receive dependences of energy oscillators on time, shown in figure 18. It is supposed, that energies of all oscillators at $t = 0$ are identical, and the phase of oscillations in the first oscillator lags behind by $\pi/2$ from phases of others oscillators. As follows from figure 18, the maximal energy in the first oscillator increases by 2,5 times, and the period of beatings decreases by 2 times as compared with figure 17. Speed of increase in energy in the first oscillator at $t = 0$ in figure 18 is greater by 8.9 times, than that in figure 17. It means, that action of 4 adjacent oscillators with favorable phases is stronger by 8.9 times than action of one oscillator

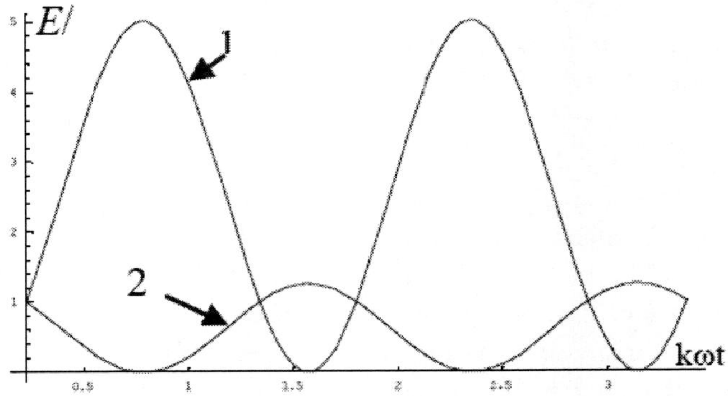

Figure 18. Dependence of energy in 5 interacting oscillators on time: 1 – energy of oscillator surrounded by 4 other oscillators; 2 – energy of each of these oscillators

Possibly, it is difficult enough to provide optimal conditions for realization of accumulation from K oscillators and they can be realized only at extremely favorable phase relations. For example, if to admit that the probability of extremely favorable phase relations between two adjacent oscillators is equal p = 0,1 instead of p = 0,5 the probability of extremely favorable phase relations in a system from $K + 1$ oscillators is equal 10^{-K}. Thus, the greater K, the more favorable conditions can be obtained, but a probability of such conditions decreases with growth K.

Usually, when any transient processes in some system are considered, a time constant is introduced which characterizes how quickly the system tends to its steady state. From this reasoning ought to introduce a time constant τ_R, which characterizes the time required to a system from a gas of densely packed oscillators with random phases to pass in the steady state when phases of all oscillators are identical.

However, before that, it is necessary to find some mechanism responsible for dissipation of energy in a non-equilibrium system because energy of the system at steady state should be minimal. Action of this mechanism determines value of the time constant. For example, in gases, this constant differs significantly for different mechanisms. In gases it is necessary only 1-2 collisions between molecules for a forward-forward relaxation, 4-5 collisions for rotary-rotary one and about 10^{20} collisions for oscillating-oscillating one [17].

In accordance with (16) own frequency of a system of two interconnected oscillators with synchronous oscillations decreases at nearing oscillators. Then in accordance with Manley-Rowe relations, energy of such system decreases. It means that oscillators are drawn one to other. As a result, in liquid or solid, consisting from such oscillators, there are forces of compression that leads to appearance of elastic fluctuations with corresponding dissipation of energy. Unfortunately, the physical nature of these forces of an attraction is not known to us. We can remind only that the physical nature of gravitation forces is not known also. Possibly, these forces can have an identical origin.

As follows from presented theoretical consideration the following conditions should be provided for noticeable increase in R. First, compression of gas should be so much strong that molecules of gas have appeared densely packed close together. Secondly, they should remain densely packed for a long time. In the third, duration of transient on compression of gas should be no more τ_R. Summarization of all this three conditions enables to claim that a quickly prepared long-living gas of densely the packed molecules is necessary for obtaining areas where a noticeable increase in R is possible

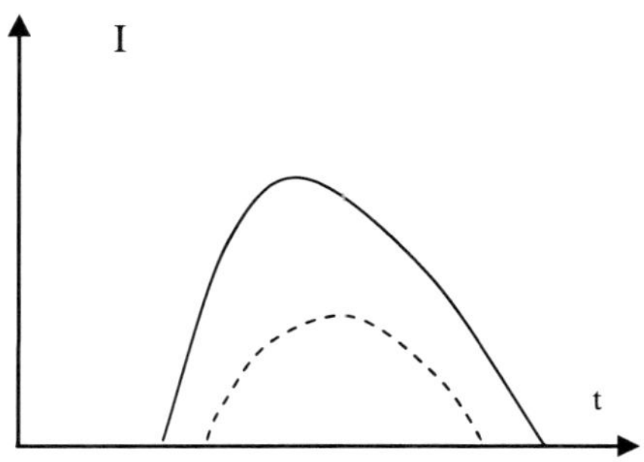

Figure 19. Dependence of the speed of radioactive decay on time. Solid curve shows corresponds to instant compression, dotted curve – gradual compression

As the maximal coupling index k_{max} between oscillators is small, areas with noticeable increase in R can be obtained if time of interaction between oscillators is greater than a certain bottom threshold τ_{min}. As follows from (21),

$$\tau_{min} = R/(2\omega k_{max}). \tag{22}$$

If duration of interaction is smaller than τ_{min}, appearance of areas with noticeable increase in R is impossible.

The solid curve in figure 19 shows dependence of the number of oscillators I which energy decreases down till E_{th} at the assumption that gas is compressed instantly at $t = 0$. The curve is received on the basis of following qualitative consideration. Time τ_1 is minimal time, which is required to accumulate the action at an assumption that the oscillator occurs at optimum phase relations with other interacting oscillators. As the probability of such situation is small, the number is close to zero. With increase in time of interaction E_{th} is reached at phase relations which differ from optimal ones a little. The probability of such phase relations is greater and, hence, the number is also greater. At the further increase in time of interaction the number I can start to decrease, because there is an alignment of phases in adjacent oscillators, and the exchange of energy between them ceases.

If compression of gas occurs not instantly, dependence of the number I on time is described by curve shown in figure 19 by dashed line. The following features of this dependence ought to notice. The delay τ_1 after which I becomes different from zero increases because of transients which take place during compression of gas. These transients at which phases of all oscillators are leveled, do not allow to receive optimum phase relations at $t > \tau_1$. The maximum of a curve decreases. Ought to underline, that I is always equal to zero at $t < \tau_1$ and, hence, long compression always is required. If speed of transients is greater than speed of compression, I is close to zero as the system is in quasi steady state where the exchange of energy between oscillators practically is absent.

Thus we can wait a noticeable increase in a speed of radioactive decay in a quickly prepared long living gas of densely packed molecules.

THE ANALYSIS OF THE PHYSICAL CONDITIONS IN EXPERIMENTS ACCOMPANIED BY NUCLEAR CONVERSIONS

Meaning simple theoretical conclusions received above, we shall analyze conditions of successful experiments known to us in which anomalous phenomena were observed. It has appeared that in all experiments quickly prepared long living gas of densely packed molecules took place. We ought to note that the shell of the light bubble consisting of strongly compressed air can appear extremely quickly during a fraction of micro second. There are some experimental confirmations of this fact [18]. It is no wonder because light participates in its formation and makes thousand revolutions in the shell in this time. Thus, physical conditions inside of an AO shell are suitable for increasing radioactive decay. AO shell is, probably, a single object among known ones which provides necessary physical conditions at which time of interaction $t > \tau_1$. Analysis of experimental studies where mention about nuclear conversions takes place shows that AOs are observed in the great bulk of these studies. There is one group of the studies where nuclear conversions are observed but direct mentions about AOs absent. But a quickly prepared long living gas of densely packed molecules is presented in these studies too. Consider results of some experiments.

At first sight, no such gas existed in classical Pons and Fleishman experiments connected with electrolysis of palladium electrodes in heavy water. Really, it seemed so within 14 years. In 2003, as a result 14-years research the following facts [19] have been established. First, areas in which nuclear activity takes place lay on a surface of the cathode, instead of inside of metal as it was supposed earlier. Secondly, these areas exist in dendrites or nano crystals, located on a surface of an electrode, instead of on the very surface. A conclusion has been drawn that the only thing necessary for a nuclear conversion is existence of the small isolated domains of a material. It can be dendrites growing on a surface of the electrode or impregnations in the palladium electrode, or grains of other material on the palladium electrode. Erosive discharges on these inhomogeneities can occur

These conclusions were confirmed in Szpak work [20], where was shown that nuclear conversion is accompanied by occurrence on a surface of electrode of "hot spots" and micro explosions. It was estimated that each micro explosion is accompanied by a nuclear conversion in which from 10^4 up to 10^9 nuclei participate. It is not excluded, that the reason of micro-explosion is disappearance of a miniature ball lightnings have appeared as a result of local erosive discharges. The term “miniature ball lightning” means the same that our tem “autonomous object”. We shall use already accepted term though the term miniature ball lightning is more often used. Japanese investigator Matsumoto [21, 22, 23] repeatedly informed about observation of AOs in conditions of classical Pons and Fleishman experiments

Lewis analyzed surfaces of those cathodes which have been used in successful attempts to carry out nuclear conversion. It turns out that these cathodes, unlike other cathodes, have been undergone to strong erosion [24]. Besides, occurrence of AOs from their surface was observed in a process of nuclear conversion. Similar erosion of electrodes is observed in numerous experiments on AO production by means of erosive gas discharge [25]. Thus, at present time it is authentically established that nuclear conversion in conditions of classical experiment is accompanied by occurrence AOs and, therefore, there is a quickly prepared long living gas of densely packed molecules.

Such gas presents also in cavitation bubbles produced in liquids by means of an intensive acoustic wave. It is known, that pressure in such bubbles at their compression can reach several thousand atmospheres. As was shown in Kladov experiences carried out in 1998-2001, nuclear conversion took place in such bubbles [26, 27, 28]. Analogous experiences have been carried out by Taleyarkhan group in 2004 [29].

Now this group is financed by DARPA. The conclusion that the gas obtained at explosions accompanied by erosion of a material is favorable for nuclear conversion has been confirmed by Ukrainian scientists who used a powerful pulse of electrons with energy about 1 kJ for super compression of substance [30].

It would seem, the gas is absent in Correa experiments connected with production of excess energy by means of vacuum discharges. [31]. However works of other researchers, in particular, Shoulders showed that the vacuum discharge is accompanied by occurrence of so called exotic vacuum objects and transmutation of elements [32]. In reality such objects represent AOs, and their shape where intensive light circulates consists of vapors of metal which has evaporated from the cathode of a vacuum tube. Thus, the gas consists of molecules of metal in this case of. The same is valid for Solin experiments with welding of zirconium in vacuum by electronic beam [33].

By the way, Sholders studies allow us to explain the existence of ectons which are avalanches of electrons arising at the vacuum discharge [34]. An appearance of an avalanche can explained somehow within a frame of usual representations without attraction of AO but it is impossible to find out reasonable reasons caused a disappearance of an avalanche. As was shown in chapter 2, introduction of AOs into consideration of the phenomena at vacuum discharges enables to explain both occurrence, and disappearance of avalanches, and also the fact of their existence. Introduction of AOs into consideration permits also to explain occurrence of excess energy in Correa experiences because any appearance of AOs is accompanied by appearance of a quickly prepared long living gas of densely packed molecules. The excess energy appears due to acceleration of nuclear decay. Indeed, transmutation of elements has been fixed in Sholders experiments.

R.F. Avramenko has informed about generation of excess energy by means of AOs in a form of so-called great energy plasma formations [25]. In this case AOs were produced by erosive gas discharge in air at normal atmosphere pressure. Urutskoev has informed about a transmutation of elements at explosions of titanic wires at discharge of the battery of capacitors about 50 kJ stored energy [18]. He has marked that the explosions are accompanied by occurrence of "spherical plasma formations". In reality, they are AOs. By the way, the fact of extremely fast occurrence of light bubbles in has been fixed in these experiments in the first time. It has appeared, that at shooting by high-speed video camera a light bubble appears so quickly, that it is absent on one frame, and it has been generated completely on the following frame.

Thus, introduction of AOs into consideration allows finding out some common and explaining seemingly absolutely different experiments. All known experiments on successful realization of nuclear conversions, where either generation of excess energy, or transmutation of elements takes place, are accompanied by an appearance of a quickly prepared long living gas of densely packed molecules.

Some views on order of magnitude constants used above can be obtained from analysis of physical conditions of successful experiments accompanied by nuclear conversions. Since life time of AO is in the interval from 1mks to 1 ms and the lifetime is comparable with the time constant for passing a system of densely packed oscillators in a steady state, then τ_{NC} is in an interval from 10^{-6} up to 10^{-3} s. The beating period is greater by 1-2 orders of magnitude than τ_{NC} and, hence, and τ_b is in an interval from 10^{-5} up to 10^{-1}s.

According to modern representations of nuclear physics "in a nucleus, considered in a form of a drop [35], there are oscillations with the period $\tau_n = 10^{-21}$s and amplitude equaled to 0,1-0,2 radius of a nucleus ". Taking into account that $\tau_n \approx 10^{-21}$ s, we obtain from (20) $k = \tau_n/\tau_b \approx 10^{-20}–10^{-16}$. Certainly, these evaluations are only an illustration of how extremely small coupling index can lead to a appreciable result for reasonable time.

Observation of appearing AOs testifies that physical conditions in AOs are favorable for acceleration of nuclear decay. Since the nature of AOs is known, these conditions can be created in laboratory. It is known, that experiments with nuclear conversions are characterized weak reproducibility. Low reproducibility takes place also in experiments on AOs production because a physical nature of AOs was not known for experimenters. Purposeful use of erosive gas discharge allows to eliminate the specified lack and to start purposeful accomplishment of nuclear conversions. In accordance with one of hypotheses, the Ball Lightning was a reason of Chernobyl tragedy [36], which essentially delayed development of atomic engineering. It is not excluded, that the Ball Lightning will expiate the fault and will play solving positive role in development of absolutely new direction in nuclear energetic which will lead to creation of new cheap alternative energy sources. How the life on the earth will be changed is described by American futurologist Arthur Clark [37].

Possible Types of Nuclear Decays

From the presented consideration follows that interaction of only identical nuclei can accelerate a nuclear decay. Besides, the smaller the difference between the energy stored in a nucleus before decay and after decay the easier conditions for realization of such conversion can be created. What types of nuclear conversions can be carried out by accumulation of interaction between adjacent identical nuclei? It is not difficult to find out a majority of possible decays in which the minimal excess energy takes place. In this case the total number of protons and neutrons is kept after conversion.

Similarly to classical reaction of synthesis of deuterium nuclei

$$2\,{}^{2}_{1}D \rightarrow {}^{4}_{2}He + 23.8 \text{ MeV},$$

We can write, for example, the following reactions

$$2\,{}^{27}_{13}Al \rightarrow {}^{26}_{12}Mg + {}^{28}_{14}Si + 3.315 \text{ MeV},$$

$$2\,{}^{63}_{29}Cu \rightarrow {}^{64}_{30}Zn + {}^{62}_{28}Ni + 1.59 \text{ MeV}.$$

The excess energy is determined by the difference in mass defects ΔM between initial and resulting products. In accordance with [38] ΔM for considered nuclei are the following:

ΔM for ${}^{27}_{13}Al$ – 17.194 MeV,

ΔM for ${}^{26}_{12}Mg$ – 16.212 MeV,

ΔM for ${}^{28}_{14}Si$ – 21.491 MeV,

ΔM for ${}^{63}_{29}Cu$ – 65.578 MeV,

ΔM for ${}^{64}_{30}Zn$ – 66.001 MeV,

ΔM for ${}^{62}_{28}Ni$ – 66.745 MeV.

Notice, that the left hand side in the record of reactions begins with number 2. It means that identical nuclei can participate in reaction only. The first reaction with deuterium is reaction of synthesis. However it is impossible to tell about the second and third reactions as before and after reaction 2 nuclei exist. After the analysis ΔM for some heavy nuclei, it is possible to write, for example, the following reaction:

$$2{}^{200}_{80}Hg \rightarrow {}^{200}_{80}Hg + 2{}^{100}_{40}Zr + 100\,МэВ$$

This is typical reaction of radioactive decay with generation 100 MeV additional energy. Actually there is a splitting Hg nucleus on 2 identical *Zr* nuclei by means of other *Hg* nucleus that is accompanied by generation of superfluous energy. Certainly, splitting can occur and on not identical nuclei. Thus, there can be a synthesis of nuclei, their splitting, and also transmutation of elements, at which 2 identical nuclei is transformed in various nuclei. Thousands of similar reactions can be imaged. The considered examples are only an illustration. The analysis of possible types of nuclear conversion is a problem of nuclear physics. We would like to note only one obvious conclusion. The smaller life time of a nucleus is the easier to carry out nuclear conversion with such nucleus. Indeed, in this case there are either relatively great radiating losses or/and a small difference E_0-E_{th}. An isotopic conversion is accompanied by small excess energy. What is why a change of a natural isotopic ratio is mentioned often.

It is very pictorial experiments of Urutskoev at explosion of a titanic wire which were accompanied by appearance of AOs. Natural isotopic ratio changed after the explosion as follows. A fraction of isotope Ti^{48} decreased from 72 % up to 62 %, percent of isotopes Ti^{46}, Ti^{47}, Ti^{49}, Ti^{50} increased from 8 % up to 10 %, from 6 % up to 8 %, from 10 % up to 12 %, from 6 % up to 8 %, respectively. Unfortunately, the paper with these data disappeared from Internet. Mention about these reactions is in [40]. Apparently, there were nuclear conversions of the following type

$$2{}^{48}_{22}Ti \rightarrow {}^{49}_{22}Ti + {}^{47}_{22}Ti + 3.485\,МэВ$$

Or

$$2^{48}_{22}Ti \rightarrow ^{50}_{22}Ti + ^{46}_{22}Ti + 0.420\ M\text{э}B$$

In the first case one neutron passes from one nucleus in others and in the second case two neutrons pass.

Probably, similar reactions use colonies microbiological cultures in radioactive waste with a rather small lifetime. There is evidence that in such environments transmutation of elements is possible. Besides noticeably increases the speed of radioactive decay [40]. Thus there are neither explosions, nor intensive light, nor any AOs.

"Micro installation" in a form of a bacterium can be imagined which have learned to break a stationary phase relation between adjacent oscillators by sharp relocation of oscillators. As a result, the coupling index between oscillators and their own frequencies change. It causes phase shift of oscillations in these oscillators relative a phase of oscillations in an isolated oscillator. When the phase shift reaches $\pi/2$, the bacterium nears the oscillator with other oscillator having the steady state phase. As a result, the phase shift between these oscillators becomes equal to $\pi/2$. This provides transmission of energy from one oscillator to others. Thus the bacterium accelerates a decrease in energy of nucleus until its energy achieves threshold E_{th} and the nucleus breaks up. Excess energy generated at this process is used by bacterium for ability to live.

As a rule, coupling index between usual nuclei is so small and the difference E_0-E_{th} is so great that the bacterium is not able to provide a necessary decrease in E_0 down to E_{th}. But the situation can be favorable for nuclei with rather small half-life period. There is double usefulness from small half-life period of a nucleus. Firstly, the smaller half-life period of a nucleus the greater the amplitude of wave function outside the nuclei because in accordance with generally accepted notions the square of the wave function outside a nucleus is proportional to the probability of its decay. In this case coupling index between adjacent nuclei is greater than that between nuclei with great half-life period. Secondly, the difference E_0-E_{th} is small enough because the excess energy is small. Radioactive waste with rather small life time contains such nuclei. Bacteria have learned to extract and use excess energy for their life earlier, than the people have. Bacteria have advantage over people because there are no theorists among bacteria who assert that it is impossible.

Certainly, it is very hard work to find out the most suitable new fuel for new alternative sources of energy. On the one hand, it should be a wide spread element with relatively long life so time that it is not dangerous for health. On the other hand, its excess energy at accelerated nuclear decay should be considerable and conditions for production of a quickly prepared long living gas of densely packed molecules should be as simple as possible. Thus, instead of a single nuclear reaction of synthesis of deuterium nuclei which takes place in H-bomb there are a lot of other nuclear reaction which are also accompanied by appearance of excess energy. Possibly, presented above our rough consideration can determine a direction of further investigations. Ought to underline there is no mystics. Indeed, natural radioactive decay is an objective reality. Possibility of changing the speed of radioactive decay is also objective reality. Appearance of excess energy and transmutation of elements are also objective realities. Everything that is required it is to find out a way of using these realities in practical applications.

CONCLUSION

Extension of processes and the phenomena occurring in a light bubble on nuclei of elements allows us to explain the phenomenon of natural radioactive decay. Just as the energy reserved in a light bubble is gradually decreasing and the light bubble becomes instable, energy of a nucleus gradually decreases in time too; the nucleus becomes instable and breaks up. It is shown that such assumption does not contradict reliable experimental data according to which average energy of a nucleus does not depend on time if interaction between nuclei is taken into account. The experimental data confirming these conclusions and testifying that external conditions can influence the speed of a radioactive decay are presented.

An attempt to find the means allowing one to increase speed of radioactive decay has been undertaken. With this purpose the nucleus has been presented in a form of nonlinear oscillator with a definite lifetime. An interaction of a set of such identical oscillators has been considered. It is shown, that for a noticeable decrease in energy of the some oscillator it is necessary to randomize initially phases of all oscillators, and then to pack these oscillators as quickly and densely as possible to provide the condition in which the relative positioning of oscillators would be kept a relatively long time. The derived conclusions can be used not only for realization of nuclear reactions of decay but also for nuclear reactions at which transmutation of elements occurs. It is shown that such reactions could be carried out in the quickly prepared long-living gas of densely packed molecules. Numerous experimental data supporting this conclusion are presented.

REFERENCES

[1] Spillane, SM; Kippenberg, TJ; Vahala, KJ. *Nature,* 2002, 415, 621–623.

[2] *Vihman E. Quantum Physics*; McGraw-Hill Book Co.: 1967.

[3] Torchigin, VP. *Docl. Phys.* 2003, 48, no. 3, 108–111.

[4] Torchigin, VP; Torchigin, AV. *Phys. Lett. A,* 2005, 337, 112–120.

[5] Torchigin, VP; Torchigin, AV. *Europhys. J. D,* 2005, 32, 383–389.

[6] *Oraevskiy, AN. Quantum electron.,* 2002, 32, № 5, 377–400.

[7] *Encyclopedia of Physics* (Mc Graw-Hill, 1993).

[8] Audi, G; Wapstra, AH. *Nucl. Phys.*, 1995 A, 595, 409. C:\Documents and Settings\Lord\Desktop\Lightning Properties, Formation and Types\page proofs\SU ITEpublished\About_Chernobyl_Tragedy.pdf

[9] *Urutskoev, LI;* Fillipov, DN. *Usp. Fiz. Nauk*; 2004, 174, № 12, 1355–1358.

[10] Jung, M. et al., *Phys. Rev. Lett.* 1992, 69, 2164.

[11] *Jelepov, BS; Ziryanova, LN. Influence of electrical field of atom on beta decay; Nauka: Moscow, Leningrad*, 1956.

[12] *Jelepov, BS; Ziryanova, LN; Suslov, Yu, P. Beta processes: functions foe analysis of beta spectrums and electron capture*; Nauka: Leningrad, 1972.

[13] *Aculov, Yu. A; Mamirin, BA. Usp. Fiz. Nauk,* 2003, 173, 1187.

[14] *Mamirin, BA; Akulov, Yu. A. Usp. Fiz. Nauk,* 2004, 174, 791.

[15] Goddard, G; Dash, O. *Trans. American Nuclear Soc.*, 2000, 83, 301.

[16] Haus, HA. *Waves and Fields in Optoelectronics*; Prentice-Hall, NJ, 1984.

[17] *Luybitov, Yu. N.* In: *Physical encyclopedia*, Prokhorov, A., M; Ed; Bolshaya Rosiyskaya Entsiklopediya: Moscow, 1998, Vol. 1, 375–379.
[18] *Urutskoev, LI. Lomonosov, 2002, 10, 8-12 (In Russian).*
[19] Storms, E. *Tenth International Conference on Cold Fusion*, USA, Massachusetts, Cambridge, August 2003.
[20] Szpak, S; Mosier-Boss, PA; Dea, J; Gordon, F. *Tenth International Conference on Cold Fusion*, USA, Massachusetts, Cambridge, August, 2003.
[21] Matsumoto, T. *Fifth International Conference on Cold Fusion*. 1995 April 9–13, Monte Carlo, Monako.
[22] Matsumoto, T. *Fusion Technology*, 1992, 22, 281.
[23] Matsumoto, T. *IEEE International Pulsed Power Conference*, 2001, 1, 273–276.
[24] Lewis, EH. In *Tenth International Conference on Cold Fusion*, USA, Massachusetts, Cambridge, August, 2003.
[25] *Avramenko, RF; Nikolaeva, VI; Poskacheva, LP.* In book *Ball lightning in laboratory*. Editor *Avramenko R. F.* Himiya, Moscow, 1994, 7-56 (in Russian).
[26] Kladov, A. In *13-th Radiochemical Conference*. 19–24 April, 1998. Marianske Lazne Jachymov Czech Republic. Booklet of Abstracts.
[27] Kladov, A. In *5-th International Conference on Nuclear and Radiochemistry*. Pontresina, Switzerland, 3–8 September, 2000, Extended Abstracts vol. 1.
[28] Kladov, A. In *21st International Symposium «Industrial Texicology 2001»*. Proceedings. 30 May – 1 June 2001, Bratislava, Slovak Republic.
[29] Taleyarkhan, RP. et al *Journal of Power and Energy,* 2004, 218(5), 345–364.
[30] Adamenko, SV; Adamenko, AS; Vysotskii, VI. *Infinite Energy,* 2004, 9 (54), 23–30.
[31] Carrel, Mike. In *Infinite Energy Magazine* 1996, Special Selection 62–70. www.infinite-energy.
[32] Shoulders, K. *Infinite Energy*, 2005, 61.
[33] Solin, MI. *Physical Thought in Russia*; 2001, no.1, 43–58.
[34] *Mesyats, GA. Ectons in vacuum discharge: discharge, spark*, arc; Nauka: Moscow, 2000.
[35] Pik-Pichak, G. A. In *Physical encyclopedia*, Prokhorov, A., M; Ed; Bolshaya Rosiyskaya Entsiklopediya: Moscow, 1990, Vol. 2, pp 238–239.
[36] *Torchigin, VP. Bulletin on nuclear energy*; 2006 № 4, 89–92.
[37] Clarke, AC. *Infinite Energy Magazine Special Selection*, 1992, pp.8–10. www.infinite-energy.
[38] Grigoriev, IS. *Handbook of Physical Quantities*, Ed. (Energoatomizdat: Moscow
[39] *Vysotskii, VI; Shevelev, VN; Tashirev, AB; Kornilova, AA. Tenth International Conference on Cold Fusion*, USA, Massachusetts, Cambridge, August, 2003.
[40] *Urutskoev, LI. Lomonosov, 2002, 10, 8-12 (In Russian).*

CONCLUSION

We can say that Ball Lightning is trying to inform people about the existence of absolutely new objects and the phenomena about which nobody earlier knew or even suspected. These are light bubbles of various kinds whose size, lifetime and stored energy can be changed over a wide range. A shell of such light bubbles can be the most varied also. It can consist of compressed air, the compressed gas whose refractive index is greater than that of surrounding space, a double-charged layer. The shell of a light bubble can also confine negative charges.

A series of new physical effects and phenomena has been derived from the mere fact of light bubble existence and analysis of their properties obtained as a result of two centuries of experimental investigations. These discoveries are valuable in their own right and can be used in other practical applications.

It cannot be ruled out that physical conditions inside of a light bubble are favorable for acceleration of radioactive decay. Therefore the further studying of light bubble properties can have a great practical application in the development of alternative energy sources on the basis of the energy that accompanies radioactive decay. It is the right of readers to judge how close the presented approach is to reality. In our opinion the approach is deserving of attention because it enables us to explain too many anomalous phenomena. It concerns the behaviour of both Ball Lightnings and autonomous objects.

Such puzzles of Ball Lightning behaviour as penetration through windowpanes, movement in a direction which does not coincide with wind direction, accompanying of airplanes and penetration into their salons, and radiation of "cold" white light can be explained only on the basis of the considered concept. Numerous anomalies in the behaviour of autonomous objects, such as jumping from a sheet of paper, burning a sheet of foil, galloping on a surface of a table with the sudden termination of a luminescence in any point of a trajectory, behaviour near a vessel with liquid nitrogen and in ultrasonic stream, and penetration outside through a hermetically sealed glass tube can be explained only on the basis of the offered concept. It gives confidence to the authors of the correctness of the chosen approach.

The theory of light bubbles presented here disproves arguments of opponents that existence of light bubbles contradicts laws of physics. Time is required for general recognition of the presented approach. Copernicus's theory required about a century for general recognition. We hope that less time will be required in our case.

As to the relationship between Ball Lightning on the one hand and excess energy and transmutation of elements on the other hand, such a relationship exists objectively as the words “Ball Lightning” are often mentioned in combinations with these words. Knowledge of physical conditions in Ball Lightning and the assumption that these conditions are favorable for acceleration of radioactive decay give much information about the processes responsible for such decay.

Methods and the means used in nonlinear optics and theory of optical resonators on the basis of waves of the whispering gallery type are transferred to the description of the phenomena inside a nucleus and inter nuclear interactions. Consideration of a nucleus as an analog of a Ball Lightning with a finite lifetime enabled us to propose a new approach to an explanation of the phenomenon of natural radioactive decay.

REFERENCES

[1] Stenhoff, M. *Ball lightning – An Unsolved Problem in Atmospheric Physics;* New York: Springer, 1999.

[2] Spillane, SM; Kippenberg, TJ; Vahala, KJ. *Nature*, 2002, 415, 621–623.

[3] Matthews, R. *New Scientist,* 200, 8, April, 23-26.

[4] Abrahamson, J; Dinnis, J. *Nature*, 2000, 403, 3, February, 519-521.

[5] Wessel-Berg, T; *Physica*, D. 2003, 182, 223-253.

[6] Haus, HA. *Waves and Fields in Optoelectronics;* Prentice-Hall, NJ, 1984.

[7] Aitchinson, JS; Al-Hemiari, K; Ironside, CN; Grant, RS; Sibbett, W. *Electron. Lett.* 1992, 28, 1879.

[8] Segev, M; Stegeman, G. *Phys. Today,* 1998, 51(8), 42.

[9] Mitchel, M; Chen, Z; Shin, M; Segev, M. *Phys. Rev. Lett.*, 1996, 77, 490.

[10] Torchigin, VP; Torchigin, AV. *Phys. Lett.* A 2005, 337, 112–120.

[11] Torchigin, VP;,Torchigin, AV. *Dokl. Phys.*, 2004, 49, No. 10,553–555.

[12] Torchigin, VP; Torchigin, AV. *Optics Comm.*, 2004, 240/4–6, 449–455.

[13] Torchigin, VP; Torchigin, AV. *Phys. Lett.*, A 2007, 361, 167–172.

[14] Torchigin, VP. *Dokl. Phys.*, 2004, 49, No. 9, 494–496.

[15] Torchigin, VP; Torchigin, AV. *Phys. Scr.* 2003, 68, 388–393.

[16] Torchigin, VP; Torchigin, AV. *Europhysics Journal D,* 2005, 36, 319–327.

[17] Torchigin, VP; Torchigin, AV. *Europhysics Journal D,* 2005, 32, 383–389.

[18] Torchigin, VP. *Dokl. Phys.*, 2003, 48, no. 3, 108–111.

In: Lightning: Properties, Formation and Types
Editor: Matthew D. Wood
ISBN: 978-1-61122-198-5

Chapter 7

AN OVERVIEW OF CLOUD-TO-GROUND LIGHTNING RESEARCH IN BRAZIL IN THE LAST TWO DECADES*

O. Pinto
Brazilian Institute of Space Research (INPE), Brazil

ABSTRACT

Brazil is the largest country in the tropical region and it is believed to have the largest cloud-to-ground (CG) lightning activity in the world, estimated from thunderstorm days and satellite data in about 50 millions CG lightning flashes every year. The first systematic observations of CG lightning in Brazil were done in the decade of 1960 through the record of the number of thunderstorm days at different sites. Later on observations of CG flashes were done by a network of flash counters, an instrumented tower and a clod-to-ground lightning location network, most of them in the Southeast region of the country. During the last two decades the research has increased considerably, expanding to other regions in the country and including the technique of high-speed cameras. The lightning location network has also expanded considerably from four sensors in 1988 to 47 sensors at present time, becoming the largest network in the tropics and the third largest national network in the world. In this overview, the main results obtained by these techniques are described and compared with observations in other locations in the world. They include approximately 100 direct current observations, 500 optical observations by high-speed cameras and more than 20 million electromagnetic field observations by lightning location technique. As a whole, they can be considered as the largest existing database on lightning in the tropical region.

* A version of this chapter was also published in *Atmospheric Science Research Progress,* edited by Chih-Hao Yang, published by Nova Science Publishers, Inc. It was submitted for appropriate modifications in an effort to encourage wider dissemination of research.

1. Brief History

The first systematic observations of cloud to ground (CG) lightning in Brazil were done in the decade of 1960 through the record of the number of thunderstorm days (thunderdays) at different sites. The observations were done up to 1980, when the first map of the CG activity in the country was obtained. However, due to the limited number of observation sites, only a very approximate map of the lightning activity was obtained (Figure 1). Based on it, it was estimated an average flash density of 7 flashes.km^{-2}.year^{-1}, indicating a larger lightning activity compared to most of the countries in the temperate region.

The thunderday observations were followed by observations carried out by a network of flash counters in part of the Southeast region of the country (Figure 2). A 43-flash counter network operated from 1985 to 1995, recording pulses by a trigger circuit with a band pass filter centered at 10 kHz. (Pinto et al., 2003a).

In 1985, an instrumented tower to record CG flashes was installed in the Morro do Cachimbo Station (MCS), in the Southeast region of the country. The station was built by the local power company, the Companhia Energética of Minas Gerais (CEMIG) under the orientation of Dr. A. J. Eriksson (Eriksson, 1979), with the goal to determine the electric parameters of direct discharges in towers. The station records the nearby CG lightning activity, the atmospheric electric field, and the direct current measurements. In addition, it provides photograph records and video images of the flashes striking the 60-m metallic tower. The tower is located on the top of a mountain about 1430 m above the sea level and 200 m above any other mountain in the region. The current is measured by two current transducers, with a bandwidth of DC to 1 MHz. The accuracy of the current and sampling time of the data were initially limited to 760 A – 1 μs and 116 A – 0.2 μs for the main and the parallel transducer, respectively. The MCS uses a fiber optic link installed into an open duct with a copper plate ground system to transmit the information from the sensor to equipment room (Pinto et al., 2005). Figures 3 and 4 show the details of the base of the tower and of the current sensors, respectively. The first results obtained at MCS were published by Trignelli et al. (1995). Later, Pinto et al. (1997) presented a comparative analysis of mean peak current at MCS with similar observations in San Salvatore, Switzerland, Italy, and South Africa, and Pinto et al. (2003b) and Visacro et al. (2004) extended the comparison to other waveform parameters.

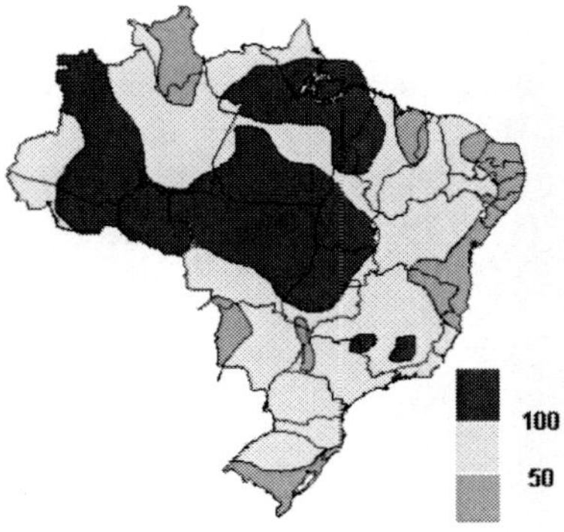

Figure 1. Map isoceraunic of Brazil obtained by thunderday observations from 1960 to 1980

Figure 2. A CIGRE 10 kHz flash counter installed in the Southeast Brazil in 1985

Figure 3. The photo shows the detail of the base of the tower in the Morro do Cachimbo Station

In 1988 the first lightning location system was installed in Brazil, also in the Southeast region (Pinto et al., 1996). The regional network was a small 4-sensor LPATS (Lightning Position and Tracking System) network. In the following years, the network expanded to cover all Southeast region and, later, other regions of the country, becoming a national network named Brazilian Lightning Detection Network (BrasilDat). Other type of sensor was added to the network, the IMPACT sensor, improving the data reliability. Also, because the network does not cover the entire country, a detailed detection efficiency model was developed to improve the data reliability mainly in the borders of the network (Naccarato et al., 2006). At the present time BrasilDat is composed by 46 sensors, covering the South, Southeast and part of the Center and North regions of the country (Figure 5), and it is the third largest network in the world and the largest in the tropics (Pinto et al., 2006a,b; 2007). The network is still growing and it is expected to cover the whole country in 2010.

Lightning observations with high-speed cameras began in Brazil in 2001. The first observations were made in Southeast region by a high-speed digital video camera model Red Lake Motion Scope 8000S (Figure 6) with a resolution and exposure time of 1 ms. All high-speed video recordings had a 1 s pre-trigger time and a total recording time of two seconds (2,000 frames). The pre-trigger time of 1 s proved to be long enough to prevent the missing of first strokes. Also, the total recording time of 2 s was long enough to capture the whole flash. All images were GPS synchronized, time stamped and without any image persistence (Saba et al., 2006). The main goal of these observations was to validate the BrasilDat data. Later, the observations were extended to the South region of the country and to investigate other lightning characteristics, in particular the flash multiplicity.

Figure 4. The photo shows the detail of the current sensors in the Morro do Cachimbo Station

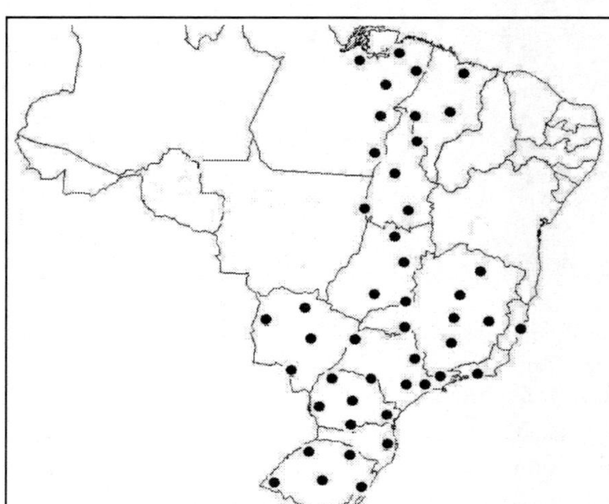

Figure 5. Map of the location of the lightning sensors of the BrasilDat at the time of this publication

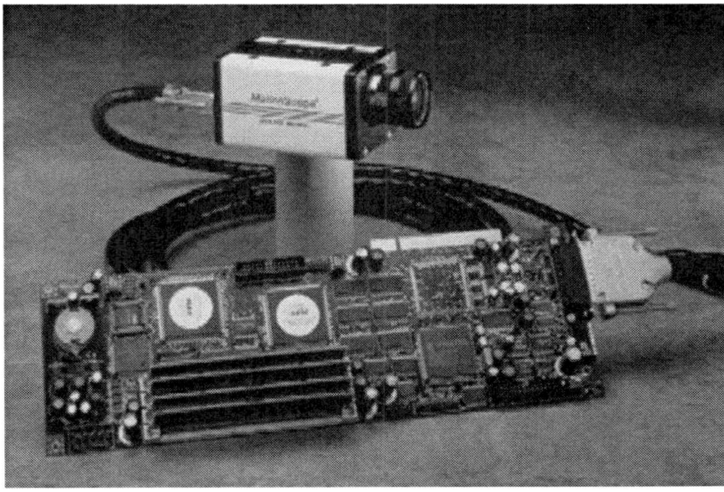

Figure 6. The photo shows the Red Lake Motion Scope 8000S high-speed camera and its PC hardware

2. Lightning Characteristics

Lightning characteristics are an important input for understanding the lightning physics, the conception of power delivery systems and to prevent lightning damages. In this section we summarize the knowledge about them obtained from observations in Brazil.

2.1. Density

The first values of flash density in Brazil were estimated from thunderstorm days. Based on these data, it was estimated an average value for the whole country of 7 flashes.km-2.year-1. More recently a annual average flash density map was published (Pinto et al., 2007) for the Southeast region of the country based on the BrasilDat network (Figure 7).

In this region of the country maximum flash densities are obtained over the city of São Paulo, suggesting an urban effect on lightning activity. For the spatial resolution of 1 km x 1 km, the maximum value in this region corresponds to 17 flashes km^{-2} $year^{-1}$, with is similar to the maximum values observed for this resolution in Europe (Schulz et al., 2005) and North America (Orville et al., 2002).

For the other regions of the country the data available from BrasilDat are still insufficient to get a statistical significant map of the lightning activity (about two years or less). At the present time, the best approach to get a map of the lightning activity for these regions is to compare the CG data obtained by BrasilDat with that the Lightning Imaging Sensor (Christian et al., 1989) in the Southeast and extrapolate the result of this comparison assuming a constant CG to intracloud flash ratio (Figure 8).

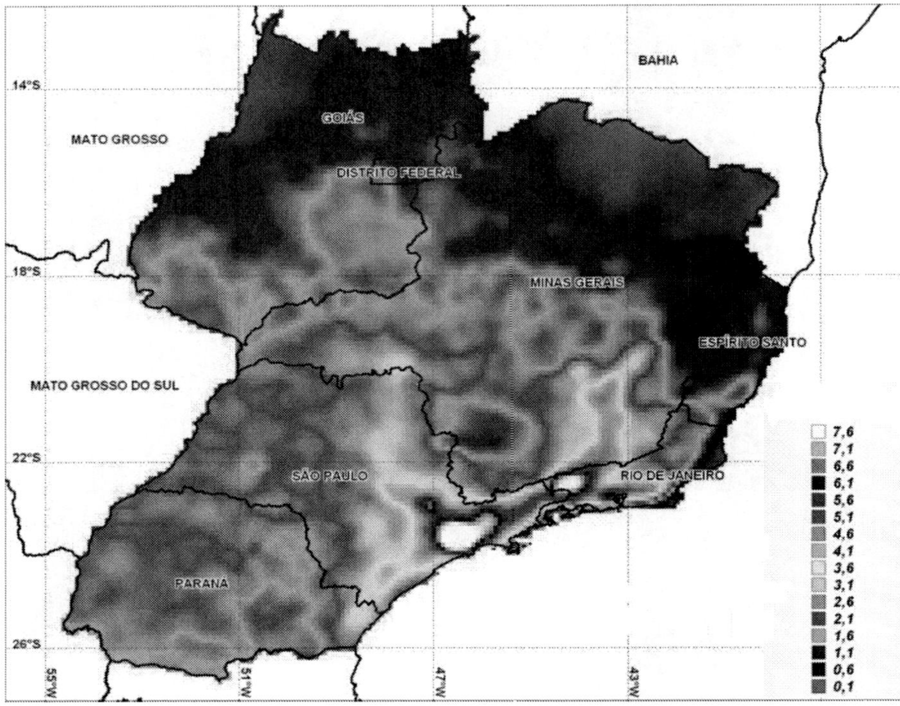

Figure 7. Annual average CG lightning flash density (in flashes.km^{-2}.year^{-1}) in the Southeastern Brazil from 1999 to 2004 for a spatial resolution of 10 km x 10 km obtained by the BrasilDat network. Regions in white correspond to densities larger than 7.5 flashes km^{-2} year^{-1}

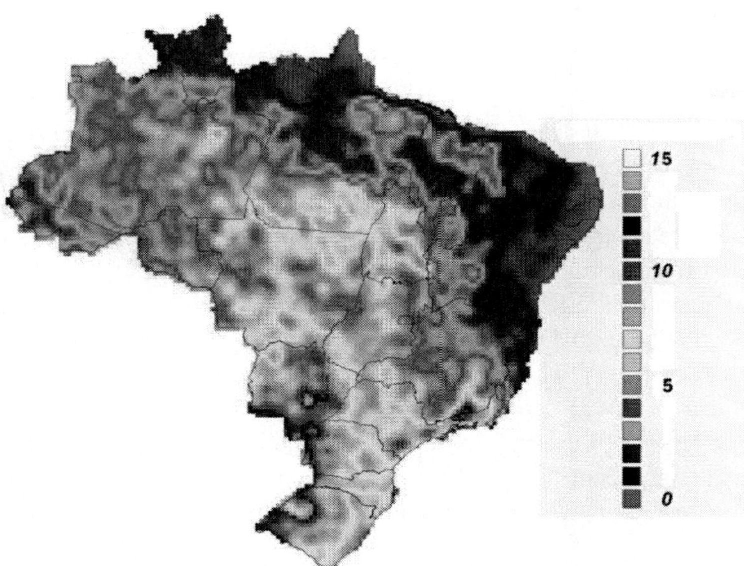

Figure 8. Map of the annual average CG lightning density (in flashes.km^{-2}.year^{-1}) in Brazil obtained from a comparison between CG lightning data from BrasilDat and the Lightning Imaging Sensor (LIS) in the Southeast of Brazil from the period 19998-2006 for a spatial resolution of 30 km x 30 km

Although this extrapolation may be not correct, mainly in the South region of the country, where large mesoscale convective systems are common and, in consequence, a large ratio should be expected, it can give more reliable information about the flash density for the whole country than that obtained from the old thunderstorm day map (Figure 1).

2.2. Polarity

The lightning polarity of CG flashes is usually expressed by the percentage of positive CG flashes. Only two sources of data are available: the MCS and BrasilDat, since the high-speed camera not allow to obtain the polarity of the flash.

The percentage of positive CG flashes at MCS is approximately 2 % (Pinto et al., 2003b), which is much lower than those observed in San Salvatore Station (SSS) in Switzerland by two similar towers, 10% and 9 % (Berger, 1967; Berger et al., 1975). Two hypotheses might be invoked to explain this difference: different latitudes or different predominant meteorological systems.

Recent results obtained in the United States by Orville and Huffines (2001) have indicated that the percentage of positive flashes depends more on the differences in the thunderstorm morphology and evolution than on the latitude.

In consequence, the observed difference in the percentage of positive flashes probably reflects a difference in the predominant thunderstorm characteristics in the tower locations.

Average distribution of the percentage of positive CG flashes obtained by BrasilDat is shown in Figure 9. In the region of the MCS the percentage obtained from BrasilDat (3-5 %) is in good agreement with that obtained from the observations in MCS.

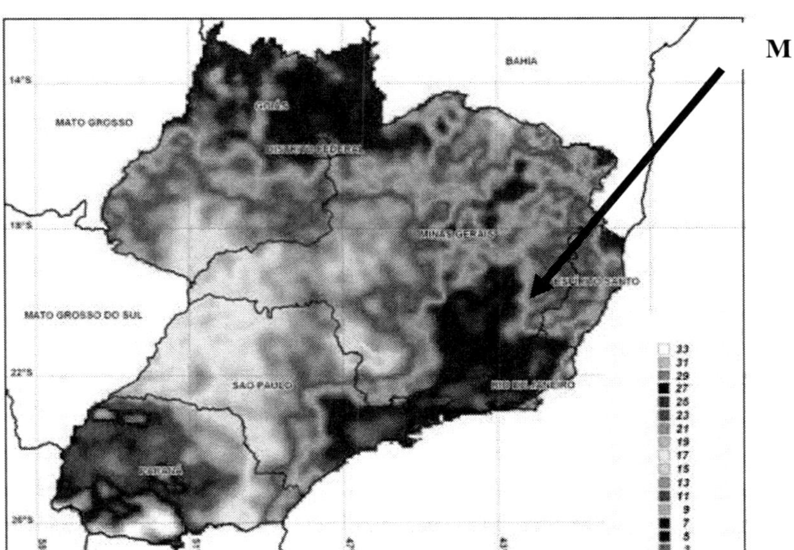

Figure 9. Average distribution of the percentage of positive CG flashes from 1999 to 2004 for a spatial resolution of 10 km x 10 km obtained in the Southeast of Brazil by BrasilDat. The location of MCS is indicated in the figure

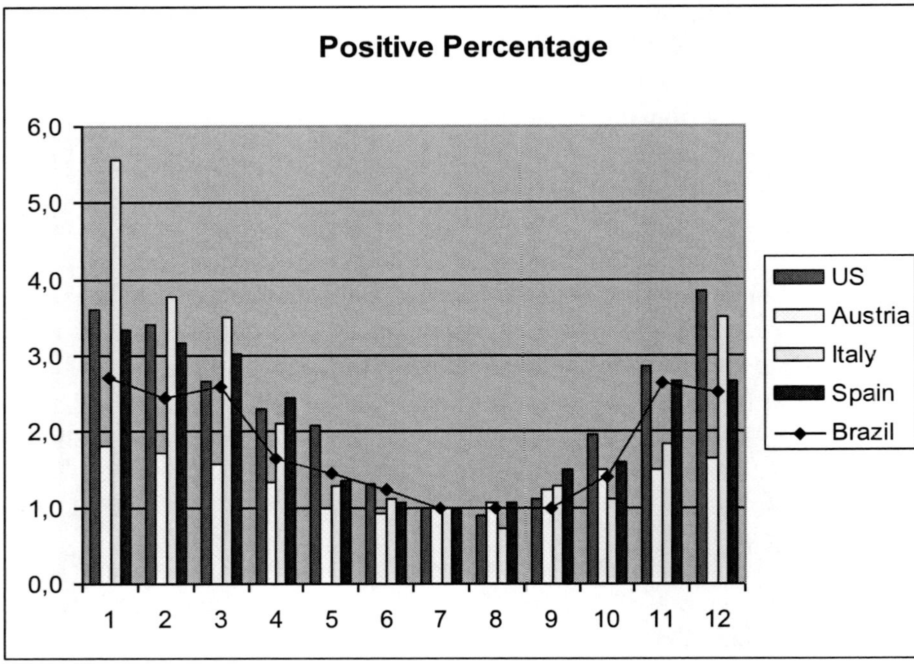

Figure 10. Normalized mean monthly distribution of the percentage of positive CG flashes observed at different countries for long time periods

From Figure 9 it can also be observed, similar to the observations by the NLDN in the United States (Orville and Huffines, 2001), the percentage of positive flashes has large variations from a few percents to almost 30%. The largest values are associated with the common occurrence of large mesoscale convective systems in the South region.

The percentage of positive flashes is also dependent on the season. Figure 10 shows the normalized mean monthly distribution of the percentage of positive CG flashes in Brazil compared to similar observations in other countries (Pinto et al., 2006b).

In order to compare them, all distributions in this figure were first shifted in time to make the months with largest lightning activity (January for Brazil, July for United States and Austria, and August for Italy and Spain) coincident. This month is referenced as month 7 in the figure. The percentage of positive CG flashes is larger in the winter for all distributions, although the variation is less pronounced in Austria.

2.3. Peak Current

Peak current of CG flashes has been measured in the MCS. The average peak current of negative CG flashes observed in the MCS is 41 kA, in contrast with the 30 kA observed in the SSS (Pinto et al., 1997). For positive CG flashes, the average peak current is not statistically significant, due to the low number of events. Peak current observations by instrumented towers like MCS are subjected to contaminations by tower reflections and biased toward higher values due to the tower ability to attract flashes increases for flashes with larger peak currents. While the influence of tower reflections on the peak current observations for towers with heights of the order of the MCS (60 m) can be ignored in a first approximation (Rakov, 2002; Guedes et al., 2003), the influence of the height of the peak current can be significant (Borghetti and Nucci, 2004). Such influence can also be enhanced if the tower is located on the top of a mountain (Rizk, 1994), as is the case in MCS that is located on the top of a 1430 m hill above sea level. As a direct consequence, the effective height of the tower is around 210 m (Erikson, 1979), what may cause a significant increase on the average peak current.

Peak currents have also been estimated from the BrasilDat network. The average values observed for negative and positive flashes are 23 kA and 28 kA, respectively (Naccarato, 2005). These values are larger than the values obtained by the North America Lightning Detection Network (NALDN), a similar network, of 17 kA and 20 kA, respectively (Orville et al., 2002). In both networks, however, large regional differences are observed. Lightning peak currents estimated by lightning location networks involve many approximations (Cummins et al., 1998). Peak current estimates from the measurements of lightning electromagnetic fields are obtained by empirical (e.g., Rakov et al., 1992; Willett et al., 1989) or theoretical (e.g., Rachidi and Thottappillil, 1993) equations relating the electromagnetic field and the lightning current. The equations, in turn, have an inherent difficulty in extracting reliable values, since they are dependent on unknown parameters such as the return stroke velocity. Although these approximations cause a significant error in the estimation of the peak current of individual flashes, for a large number of flashes the error in the average peak current is significantly reduced (Rachidi et al., 2004). In addition, the average peak current obtained by lightning location networks is very sensitive to the detection efficiency of the network. The lower is the detection efficiency, the larger is the average peak current. For instance, for the Austrian Lightning Detection and Information System, believed to have a larger detection efficiency compared to BrasilDat and NALDN, the average peak current of negative and positive CG flashes for the period from 1999 to 2001 is 13 kA and 19 kA, respectively (Schulz et al., 2005). Naccarato (2005) have developed a peak current dependent detection efficiency model, so that the average peak current can be corrected by this effect.

Considering the intrinsic limitations of the techniques discussed above, the results suggest that the average peak current of CG flashes in Brazil is not much different that observed in other countries (for a review, see Rakov and Uman, 2003). This suggestion is in general agreement with the comparison of recent observations made by a same technique in Europe and Brazil by Huntrieser et al. (2007).

Figure 11 shows the dependence of the peak current of positive and negative flashes on the season in Brazil and other countries in the same format of Figure 10 (Pinto et al., 2006b). The positive peak current has a well-defined season variation with larger values during the winter for all countries, although the variation in Brazil is less pronounced. This behaviour is less evident for the negative peak current distributions in all countries. In case of Brazil, no dependence on season is evident at all.

2.4. Multiplicity

Positive flashes are basically single stroke flashes. For this reason, the discussion is concentrated only in negative CG flashes. The percentage of multiple negative CG flashes at MCS (52 %) is higher than those at SSS for the two towers, 31 % and 35 % (Berger, 1967;

Berger et al., 1975). Two hypotheses may be invoked to explain this difference: different latitudes or different predominant meteorological systems. Recent results obtained in the United States by Orville and Huffines (2001) have indicated that the multiplicity of negative flashes has regional variations. The authors speculate that these variations may be related to variations in the horizontal dimensions of the thunderstorms at different locations. Such a hypothesis, if proved, could explain the observed difference found in this work.

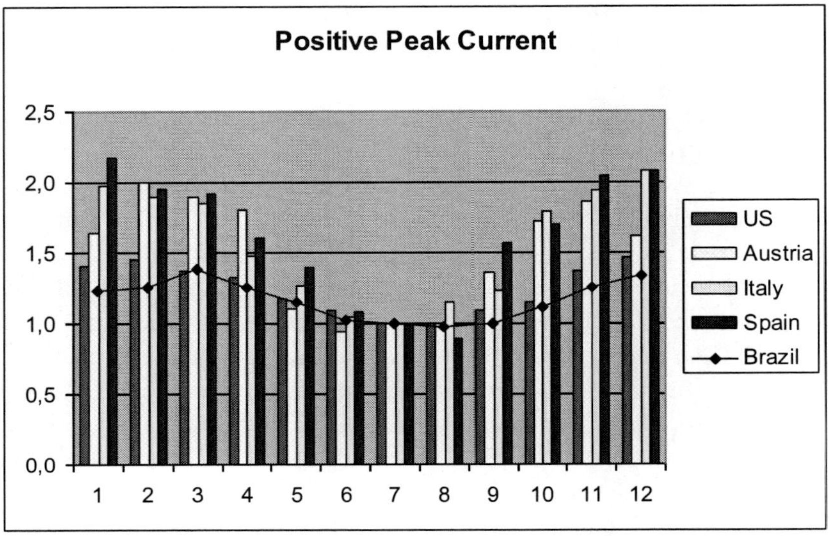

(a)

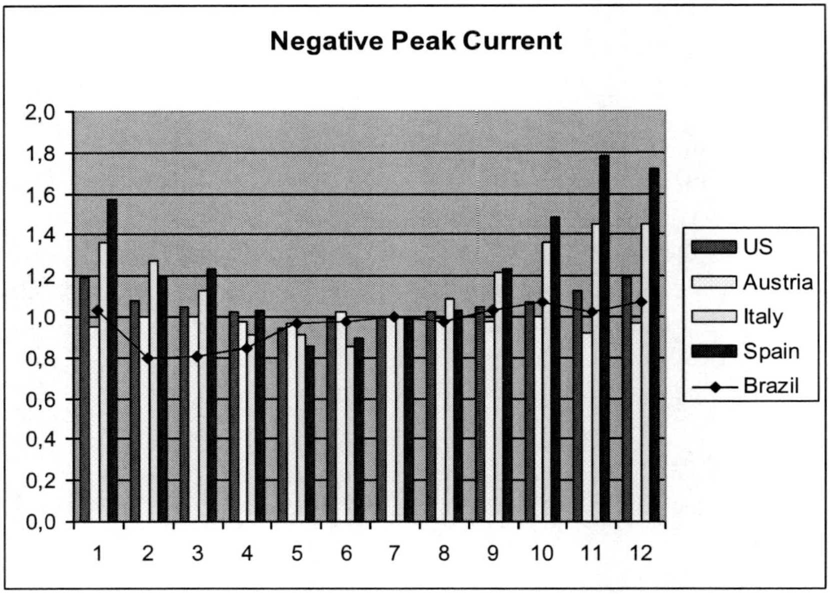

(b)

Figure 11. Normalized mean monthly distribution of the negative (a) and positive (b) peak current of CG flashes observed at different countries for long time periods

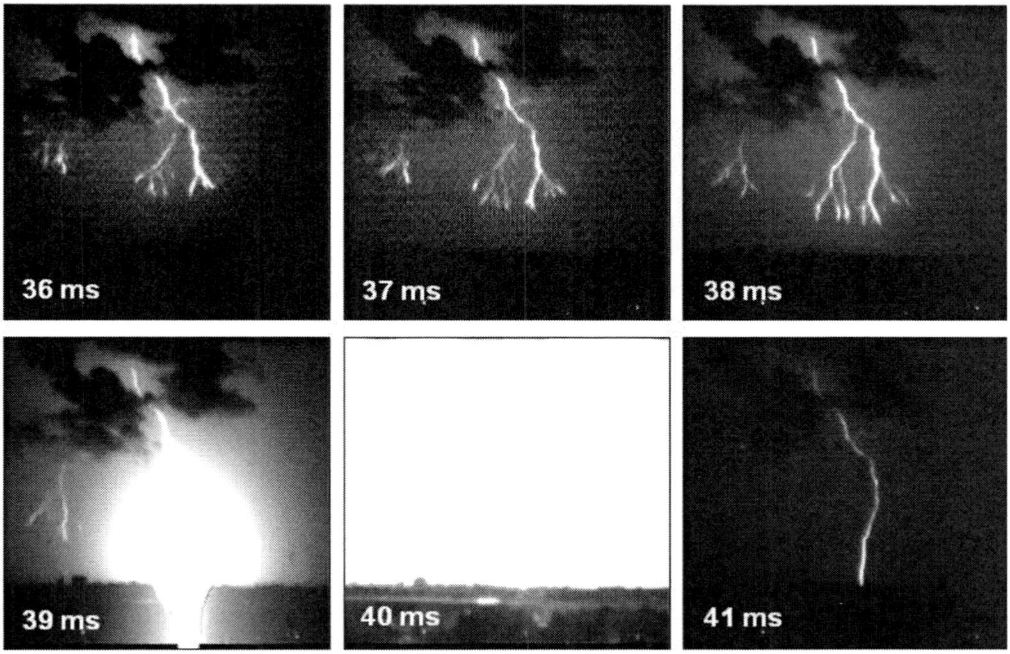

Figure 12. Example of a return stroke with its preceding stepped-leader recorded in a sequence of frames of the high-speed camera

The multiplicity of negative CG flashes has also been reported by BrasilDat (Naccarato, 2005). However, due to the low stroke detection efficiency of the network (about 55 %), the average value (1.9) is underestimated considerably. The same is observed by other networks.

The most accurate-stroke-count technique to study lightning multiplicity is the high-speed cameras. This technique enables the identification of each stroke and its termination to ground. Figure 12 shows an example of a return stroke with its preceding stepped-leader recorded in a sequence of frames of the high-speed camera. The multiplicity obtained in the Southeast region of Brazil for negative CG flashes is 3.8. This value is similar to those obtained by accurate-stroke-count techniques in other regions (Saba et al., 2006).

CONCLUSIONS

The CG lightning data obtained in Brazil in the last two decades allow quantifying for the first time the lightning characteristics in the country. Considering that the dataset is the largest ever obtained in the tropics they also can be considered as representative of the tropical region of the world. The results suggested that, except for the higher flash density, the other characteristics are similar to those obtained in the temperate region.

ACKNOWLEDGMENTS

The author would like to thank his colleagues Iara R.C.A. Pinto, Marcelo M.F. Saba and Kleber P. Naccarato for supporting and fruitful discussions, which were fundamental to achieve the results presented in this overview. The author would also like to thank Dr. Richard Orville from Texas AandM University for reviewing the manuscript.

REFERENCES

Berger, K. *J. Franklin Institute,* 1967, 283, 478-525.

Berger, K; Anderson, RB; Kroninger, H. *Study Committee n°, 33 – Cigré* 1975, 41, 3-37.

Christian, HJ; Blakeslee, RJ; Goodman, SJ. *J. Geophys. Res.,* 1989, 94, 13329-13337.

Cummins, KL; Murphy, MJ; Bardo, EA; Hiscox, WL; Pyle, RB; Pifer, AE. *J. Geophys. Res.,* 1998, 103, 9035-9044.

Eriksson, AJ. *Ph.D. Thesis* 1979, University of Natal, Pretoria, South Africa, 400.

Guedes, D; Pinto, Jr., O; Visacro, S. *Proceedings of the International Seminar on Lightning Protection (SIPDA)* 2003, Curitiba, Brazil.

Huntrieser, H; Schumann, U; Schlager, H; Holler, H; Giez, A; Betz, HD; Brunner, D; Forster, C; Pinto Jr., O; Calheiros, R. *Atmos. Chem. Phys. Discuss.,* 2007, 7, 14813-14894.

Naccarato, KP. *PhD. Thesis,* 2005, INPE, 258 p. (in Portuguese).

Naccarato, KP; Pinto, Jr., O; Pinto, IRCA. *Proceedings of the 19th International Lightning Detection Conference(ILDC),* 2006, Vaisala, Tucson, Jun.

Orville, RE; Huffines, GR. *Mon. Wea. Rev*., 2001, 129, 1179-1193.

Orville, RE; Huffines, GR; Burrows, WR; Holle, RL; Cummins, KL. *Mon. Wea. Rev.,* 2002, 130, 2098-2109.

Pinto Jr., O; Gin, RBB; Pinto, IRCA; Mendes Jr., O; Diniz, JH; Carvalho, AM. *J. Geophys. Res.,* 1996, 101, 29627-29635.

Pinto, Jr., O; Pinto, IRCA; Lacerda, M; Carvalho, AM; Diniz, JH; Cherchiglia, LCL. *J. Atmos. Solar-Terr. Phys.,* 1997, 59, 1881-1883.

Pinto, Jr., O; Pinto, IRCA; Diniz, JH; Cazetta Filho, A; Cherchiglia, LCL; Carvalho, AM. *J. Atmos. Solar-Terr. Phys.* 2003a, 65, 739-748.

Pinto Jr., O; Guedes, DG; Saba, MMF; Pinto, IRCA; Lacerda, M. *Ann. Geophysicae,* 2003b, 21, 1-5.

Pinto Jr., O; Pinto, IRCA; Saba, MMF; Solorzano, NN; Guedes, D. *Atmos. Res.,* 2005, 76, 493-502.

Pinto Jr., O; Naccarato, KP; Saba, MMF; Pinto, IRCA; Abdo, RF; Garcia, SA. de M; Cazetta Filho, A. *Proceedings of the 19th International Lightning Detection Conference (ILDC)* 2006a, Tucson, AZ.

Pinto Jr., O; Naccarato, KP; Pinto, IRCA; Fernandes, WA; Neto, OP. *Geophys. Res. Lett.* 2006b, 33, L09811, doi:10.1029/2006GL026081.

Pinto Jr., O; Pinto, IRCA; Naccarato, KP. *Atmos. Res.,* 2007, 84, 189-200.

Rachidi, F; Thottappillil, R. *J. Geophys. Res.,* 1993, 98, 18315-18320.

Rachidi, F; Bermudez, JL; Rubinstein, M; Rakov, VA. *J. Electrostatics,* 2004, 60, 121-129.

Rakov, VA; Thottappillil, R; Uman, MA. *J. Geophys. Res.,* 1992, 97, 11527-11533.

Rakov, VA; Uman, MA. *Lightning: Physics and Effects* 2003, Cambridge, Cambridge University Press.

Rakov, VA. *Proceedings of the International Conference on Lightning Physics and Effects (LPE)* 2002, Rio de Janeiro, Brazil.

Rizk, FAM. *IEEE Trans. Power Delivery,* 1994, 9, 162-193.

Saba, MMF; Ballarotti, MG; Pinto Jr., O*J. Geophys. Res.,* 2006, 111, D03101, doi:10.1029/2005JD006415.

Schulz, W; Cummins, K; Diendorfer, G; Dorninger, M. *J. Geophys. Res.,* 2005, 110, D09101, doi: 10.1029/2004JD005332.

Triginelli, WAC; Carvalho, AM; Diniz, JH; Cherchiglia, LCL. *Proceedings of the 3rd International Symposium on Lightning Protection (SIPDA),* 1995, São Paulo.

Visacro, S; Soares Jr., A; Schoroeder, MAO; Cherchiglia, LCL; Souza, VJ. *J. Geophys. Res.,* 2004, 109, D01105, doi:10.1029/2003JD003662.

Willett, JC; Bailley, JC; Idone, VP; Eybert-Berard, A; Barret, L. *J. Geophys. Res.,* 1989, 94, 13,275-13,286.

INDEX

D

E

I

J

L

M